International Series in Operations Research & Management Science

Volume 247

More information about this series at http://www.springer.com/series/6161

Andrew G. Glen • Lawrence M. Leemis
Editors

Computational Probability Applications

Springer

Editors
Andrew G. Glen
Department of Mathematics
and Computer Science
The Colorado College
Colorado Springs, CO, USA

Lawrence M. Leemis
Department of Mathematics
The College of William and Mary
Williamsburg, VA, USA

ISSN 0884-8289 ISSN 2214-7934 (electronic)
International Series in Operations Research & Management Science
ISBN 978-3-319-43315-8 ISBN 978-3-319-43317-2 (eBook)
DOI 10.1007/978-3-319-43317-2

Library of Congress Control Number: 2016960577

Printed on acid-free paper

This Springer imprint is published by Springer Nature
The registered company is Springer International Publishing AG
The registered company address is: Gewerbestrasse 11, 6330 Cham, Switzerland

Preface

In the spring of 1994 at the College of William & Mary, we started work on a project that would end up being a long-lasting source of research. We explored the idea of combining a computer algebra system (Maple V at the time) and probability results to see if the computer could be useful in performing operations on random variables and finding new distributions. Over the next 4 years, a series of procedures written in Maple started to form its own programming language, soon to be called A Probability Programming Language (APPL). Furthermore, the language and the results that the language helped produce were starting to contribute to a field of research we called computational probability. The program APPL, unlike statistical software that works on data values, is designed to work on random variables and the various functions that describe their distribution. APPL helps derive distributions of functions of random variables, probabilistic models, and other transformations. Soon after, Diane Evans joined the team and wrote procedures for discrete distributions. The two sets of procedures were put together, and in a 2001 article in *The American Statistician* [60], the launch of this open-source software began.

In 2008, John Drew, along with Evans, Glen, and Leemis, put together a monograph explaining the creation of APPL and some of the important results from the research. This book, *Computational Probability: Algorithms and Applications in the Mathematical Sciences* [46], established the state of APPL at the time, primarily how it evolved and its major algorithms. Camille Price contacted us recently and requested that we update the original monograph and write a second monograph that summarizes some more recent work. The purpose of this, the second monograph, is twofold. First, we want to combine in this one document some of the recent results that have come about with the language. Second, we want to inspire future users, professors, students, and researchers to bring APPL into their work, their classroom, and their mindset. Just as Word, Excel, LaTeX, and PowerPoint are vital yet ubiquitous elements to many researchers, we hope that APPL will become such a research tool that enables a probabilist or statistician the ability to explore new ideas, methods, and models.

Much of what is contained in the chapters that follow was published in journals over the last 20 years. Some of the works in the monograph are original efforts, yet to be published. These works highlight interesting examples, often done by undergraduate students and graduate students, that can serve as templates for future work. Each chapter is a stand-alone publication, with the authors recognized, and a short description of the importance that APPL had in the research. Furthermore, as an open-source language, it sets the foundation for future algorithms to augment the original code. Some papers heavily rely on APPL procedures; others enjoy the ease of use of data structures. Still others have added procedures to the base language.

The editors would like to thank the many people who have contributed, supported, and encouraged this effort. Each chapter author clearly has been instrumental in furthering this cause, and they are recognized at the start of each chapter. Many friends and colleagues have also been immensely supportive over the years. We would especially like to recognize the lifelong support of our wives, Jill Leemis and Lisa Glen, who have put up with our wild ideas, even though if often meant more work for them in other areas. Our children Lindsey, Mark, Logan, Andrea, Rebecca, Mary, Grace, Gabriel, Anna, Michael, and Claire have all been supportive and patient "listeners" to their fathers. Our many colleagues over the years deserve our heartfelt thanks: Richard Bell, Roger Berger, Barry Bodt, Fr. Gabriel Costa, Kevin Cumminskey, Sam Ellis, James Fritz, Ben Garlick, Grant Hartman, Steven Horton, Ted Hromadka, Michael Huber, Steven Janke, Rex Kincaid, Chris Marks, Joe Myers, Bill Pulleyblank, Tess Powers, Matthew Robinson, Mick Smith, Alex Stodala, Rod Sturdivant, Fred Tinsley, Dave Webb, Chris Weld, Joanne Whitner, and Wei Yin-Loh.

The editors gratefully acknowledge support from the Army Research Office for providing funding in their grant number 67906-MA.

Colorado Springs, CO, USA — Andrew G. Glen
Williamsburg, VA, USA — Lawrence M. Leemis

Contents

1

Accurate Estimation with One Order Statistic

Andrew G. Glen

Abstract Estimating parameters from certain survival distributions is shown to suffer little loss of accuracy in the presence of left censoring. The variance of maximum likelihood estimates (MLE) in the presence of Type II right-censoring is almost un-degraded if there also is heavy left-censoring when estimating certain parameters. In fact, if only a single data point, the rth recorded failure time, is available, the MLE estimates using the one data point are similar in variance to the estimates using all r failure points for all but the most extreme values of r. Analytic results are presented for the case of the exponential and Rayleigh distributions, to include the exact distributions of the estimators for the parameters. Simulated results are also presented for the gamma distribution. Implications in life test design, and cost savings are explained as a result. Also, computational considerations for finding analytic results, as well as simulated results in a computer algebra system, are discussed.

This paper, originally published in *Computational Statistics and Data Analysis*, Volume 54 in 2011, is an arch-typical article that relied on APPL as its palette for conducting exploratory research. Originally designed to determine how much information was lost in censoring, the article instead reports on how little information is lost as long as one knows at least one of the order statistics of a lifetest. The use of APPL's `OrderStat` procedure to derive the PDF of an order statistic is of primary importance to this paper. Furthermore, in the span of over a year, the author created dozens of Maple worksheets with APPL code that eventually resulted in this paper. APPL was used in simulations, transformations, and maximum likelihood estimation. APPL derives exact distributions of test statistics so that exact p-values were calculable.

A.G. Glen (✉)
Department of Mathematics and Computer Science, The Colorado College, Colorado Springs, CO, USA
e-mail: andrew.glen.1984@gmail.com

A.G. Glen, L.M. Leemis (eds.), *Computational Probability Applications*, International Series in Operations Research & Management Science 247,
DOI 10.1007/978-3-319-43317-2_1

Keywords Maximum likelihood • Order statistics • Type II right censoring • Computational probability

1.1 Introduction

So much is known about the exponential distribution, that finding new and useful results is difficult. While studying the effects of increasing left censorship on life test data, it was observed that very little accuracy (variance and mean square error (MSE) of the estimates) was lost as more of the "left" portion of the data was censored. So much so, that this study evolved into examining the effects of estimating parameters when the left and right censored values met and only one order statistic was observed. For various survival distributions in the case of extreme left and right censorship, it was noted that very little accuracy was lost for most order statistics. What was meant as a study to show degradation of missing observations instead reveals that little accuracy is lost due to censoring, so much so that accurate estimation with only one order statistic is possible.

For background purposes, consider the following. In a life test, in which n units are simultaneously placed on test, it is often the case that inference is needed prior to the last item failing. Right-censoring is common, often in the form of Type II right-censoring, in which the first r failure times are observed. There are many studies on inference for various types of right-censoring, to include maximum likelihood estimation of parameters. While there are many studies on right-censoring in life tests, there is remarkably little discussion of left-censoring, especially left-censoring that accompanies right-censoring. Left-censoring can occur in a number of ways. An example presented by Nelson [120] reports that the first five readings of certain contaminants were missing because the values were too small to be read by the instruments. Another example of a similar situation is presented in Leemis and Shih [97, p. 183]. Anderson and Rønn [3] consider experiments in which subjects are evaluated at only one point in their life, thus, left-censoring or right-censoring is encountered if the factor in question is present or not on the subject. Goto [62] discusses left-censoring when considering unemployment durations and proposes conditional MLE methods to handle these cases. Other articles on the subject of left-censoring include Cui [39], discussing non-parametric estimation, and Samson et al. [143], applying stochastic approximations to an HIV model. Also, left-censoring can also occur if a failure event happens when the observer is unavailable, either by accident or design. There are a number of papers on using Fisher information matrices to find asymptotic variances for estimators, see Gupta and Kundu [63] as well as Gertsbakh [58], for example. Also, Zheng and Park [178] compare the asymptotic efficiencies of MLE estimates for complete samples of size n compared to censored samples of right censored at n but from a sample size of $N > n$. This article finds exact distributions for the estimators, which result in exact variances, so Fisher matrices are not needed for asymptotic results.

1.2 The Case of the Exponential Distribution

In the case of the exponential distribution, in which only one order statistic is known, the exact distribution of the estimator is calculable and its variance is demonstrably small for most of the order statistics. Let $X_{(r)}$ be the rth order statistic from a life test with n items on test. Also, let l be the position of the first data point observed, and r be the position of the last data point observed. Furthermore, let all observations prior the lth be unknown (left-censored data), all observations after the rth be unknown (right-censored data), and all points in between be known. The probability density function (PDF) of the exponential distribution is

$$f(x) = \frac{1}{\theta} e^{-x/\theta} \qquad x, \theta > 0.$$

The general maximum likelihood function with left- and right-censored data is proportional to

$$L(\theta) = F(x_{(l)})^{(l-1)} \times \prod_{i=l}^{r} f(x_{(i)}) \times S(x_{(r)})^{(n-r)}$$

in which f, F, S correspond to the PDF, cumulative distribution function (CDF), and survivor function (SF) of the random variable X with parameter θ. In the case when only the rth data point is known, in other words $l = r$, the likelihood function reduces to

$$L(\theta) = F(x_{(r)})^{(r-1)} \times f(x_{(r)}) \times S(x_{(r)})^{(n-r)}.$$

Taking the natural logarithm and substituting the exponential PDF, CDF, and SF, the resulting log-likelihood function becomes

$$\ln L(\theta) = -\ln(\theta) - \ln\left(1 - \exp(-x_{(r)}/\theta)\right)^{r-1} + \left(x_{(r)} + nx_{(r)} - rx_{(r)}\right)\theta^{-1}.$$

The first derivative of $\ln L$ is

$$\frac{\partial \ln L(\theta)}{\partial \theta} = \frac{\exp(x_{(r)}/\theta)(-\theta + x_{(r)} + nx_{(r)} - rx_{(r)}) + \theta - nx_{(r)}}{\theta^2 \exp(x_{(r)}/\theta) - 1}.$$

Setting the first derivative equal to zero and solving for θ results in the following equation for the maximum likelihood estimator

$$\hat{\theta} = c_{n,r}\, x_{(r)},$$

for

$$c_{n,r} = \frac{-1}{\text{RootOf}(e^{z} - z - zn + zne^{z} + zr - 1)}.$$

The denominator requires finding the root of the expression for the variable z. Although the denominator of $c_{n,r}$ does not solve algebraically, it is straightforward to calculate numerically. By inspection, one (trivial) root of the denominator is zero, and further graphical analysis shows that the other real root is negative. Therefore, $c_{n,r}$ is a positive number that starts at $c_{n,1} = n$ for the first order statistic and gets smaller, approaching zero, as r approaches n. In other words, an order statistic for the exponential distribution is proportional to its mean and vice versa. Table 1.1 gives the $c_{n,r}$ values to four digits for various n and r combinations.

Table 1.1. Values of $c_{n,r}$ for various n and r values for the exponential distribution Note the value that is italicized is used in the calculations of the exponential example in Section 1.3 of the article

r	$n = 5$	$n = 10$	$n = 15$	$n = 20$	$n = 25$	$n = 30$	$n = 35$	$n = 40$	$n = 45$	$n = 50$
1	5.0000	10.0000	15.0000	20.0000	25.0000	30.0000	35.0000	40.0000	45.0000	50.0000
2	2.2314	4.7412	7.2443	9.7457	12.2466	14.7472	17.2476	19.7479	22.2481	24.7483
3	1.2907	2.9814	4.6547	6.3246	7.9931	9.6609	11.3284	12.9957	14.6629	16.3299
4	0.7987	2.0953	3.3564	4.6115	5.8643	7.1162	8.3675	9.6185	10.8693	12.1198
5	0.4671	1.5575	2.5742	3.5814	4.5855	5.5881	6.5899	7.5912	8.5922	9.5931
6		1.1924	2.0496	2.8927	3.7314	4.5681	5.4038	6.2389	7.0735	7.9079
7		0.9242	1.6718	2.3989	3.1200	3.8386	4.5558	5.2722	5.9881	6.7036
8		0.7135	1.3853	2.0267	2.6602	3.2904	3.9189	4.5465	5.1734	5.7999
9		0.5358	1.1591	1.7353	2.3012	2.8631	3.4228	3.9814	4.5393	5.0966
10		0.3673	0.9743	1.5004	2.0129	2.5203	3.0253	3.5288	4.0314	4.5335
11			0.8188	1.3062	1.7757	2.2390	2.6993	3.1579	3.6154	4.0723
12			0.6837	1.1423	1.5769	2.0037	2.4269	2.8482	3.2683	3.6876
13			0.5622	1.0014	1.4073	1.8037	2.1959	2.5857	2.9742	3.3618
14			0.4469	0.8780	1.2607	1.6314	1.9971	2.3602	2.7216	3.0821
15			0.3248	0.7682	*1.1322*	1.4812	1.8242	2.1642	2.5023	2.8393
16				0.6685	1.0182	1.3489	1.6723	1.9922	2.3100	2.6266
17				0.5761	0.9160	1.2311	1.5376	1.8399	2.1400	2.4385
18				0.4878	0.8233	1.1254	1.4171	1.7041	1.9884	2.2710
19				0.3992	0.7383	1.0298	1.3087	1.5820	1.8523	2.1208
20				0.2997	0.6593	0.9425	1.2103	1.4716	1.7295	1.9853
21					0.5847	0.8623	1.1206	1.3712	1.6179	1.8623
22					0.5131	0.7880	1.0382	1.2794	1.5161	1.7503
:						:				:
25					0.2825	0.5906	0.8254	1.0446	1.2570	1.4658
:						:				:
30						0.2698	0.5427	0.7450	0.9318	1.1118
:										:
35							0.2598	0.5069	0.6860	0.8497
:										:
40								0.2517	0.4791	0.6405
:										:
45									0.2450	0.4566
:										:
50										0.2392

Logically, the values of $c_{n,r}$ for the smaller order statistics are less than the mean of the exponential(θ) and so to estimate that mean, one multiplies the smaller order statistics by a $c_{n,r} > 1$. Then as the order statistic position approaches the median, $c_{n,r}$ tends towards 1 (but because the exponential is skewed, the constant of the median will be larger than 1). Finally, for the later order statistics, $c_{n,r}$ gets closer to zero, as the order statistics are greater than the mean. More importantly, knowing the value of $c_{n,r}$ enables one to find the exact distribution of the estimator $\hat{\theta}$ in the case of assumed exponentiality as well as exact confidence limits for θ. First consider the random variable,

$\hat{\theta} = c_{n,r}\, X_{(r)}$. One can derive the distribution of $X_{(r)}$ and find the distribution of $\hat{\theta}$ using the transformation technique from mathematical statistics to find the PDF of the estimator. Furthermore, while it is possible to do by hand, this type of exploration is made less tedious using computer algebra software. In this case the author uses A Probability Programming Language (APPL) [46] and [60] running in Maple for all of the derivations in this paper. APPL is an open source set of procedures available from the author. Also, all code in this paper is available from the author. An illustration using APPL code is provided below for finding the distribution of the estimator.

```
> assume(n::posint); # set up assumptions
> assume(r::posint);
> assume(theta > 0);
> assume(cnr > 0);
> additionally(n >= r);
> X := [[x -> exp(-x / theta) / theta],[0, infinity],
       ["Continuous", "PDF"]];
       # creates an unspecified exponential distribution with
       # parameter theta
> Xos := OrderStat(X, n, r);
       # calculates the unspecified PDF of the rth order statistic
> Thetahat := Transform(Xos, [[x -> cnr * x],[0, infinity]]);
       # calculates the PDF of the estimator
```

The result is the PDF of the estimator

$$f_{\hat{\theta}}(y|\theta) = \frac{n!\,\left(1-\exp(-y/(\theta\, c_{n,r}))\right)^{r-1} \exp(-(n-r+1)\, y/(\theta\, c_{n,r}))}{(r-1)!\,(n-r)!\,\theta\, c_{n,r}} \qquad y > 0.$$

Further, for any fixed n and r, the expected value and variance of the estimator are calculable. For example, consider the 22nd order statistic for increasing values of n = 30, 35, 40, and 50. The mean and variance of each of these estimators are calculated in APPL with the additional commands similar to

```
> Mean(subs({n = 30, r = 22, cnr = 0.78803}, Thetahat));
> Variance(subs({n = 30, r = 22, cnr = 1.038175}, Thetahat));
```

The resulting values for the mean and variance of $\hat{\theta}$ are listed in Table 1.2.

Note that we can calculate the exact bias of the estimator. Furthermore, as we would expect, because MLEs are asymptotically unbiased, we see the bias decreasing as n increases. Also shown is the effect on variance of changing n, in case an experimenter knows a desired variance threshold in advance. This type of analysis helps one design an experiment, especially choosing appropriate n and r that will produce a timely end to the experiment. Also, it was noted (but not shown in this table) that $E(c_{n,1}X_{(1)}) = \theta$, in other words the MLE using the first order statistic is unbiased (although it has extraordinarily high variance). For each subsequent order statistic there is increasing bias, so much so that about half-way through the order statistics, the bias factor is on the order of 1.004, which is still less than one-half a percent of the value of θ.

Table 1.2. Exact mean and variance for estimators using only the 22nd order statistic from various sample sizes

Order statistic	Mean	Variance
$n = 30,\ r = 22$	1.006418θ	$0.052616\theta^2$
$n = 35,\ r = 22$	1.003550θ	$0.049442\theta^2$
$n = 40,\ r = 22$	1.002293θ	$0.048040\theta^2$
$n = 50,\ r = 22$	1.001204θ	$0.046817\theta^2$

As mentioned in the introduction, the variance of the single order statistic estimators compare very favorably with the MLE estimators when there is no left censoring and all of the first r order statistics are known. Because the exact distribution of estimator for the single order statistic is known, its exact variance can be calculated. Likewise, for the estimates from Type II right censoring, it is well established that

$$\frac{2r\hat{\theta}}{\theta} \sim \chi^2_{2r},$$

(see Leemis [93, p. 199], for example) so that exact variances can be calculated for this estimator as well. Figure 1.1 displays the variances of these two estimators for various r and $n = 30$. The lower set of points is the variance of the MLE $\hat{\theta}$ if all order statistics $x_{(1)}, x_{(2)}, \ldots, x_{(r)}$ are observed. The upper set of points is the variance of the estimator if only one order statistic, $x_{(r)}$, is observed.

Note that the variances of these two estimators indicate that knowing more than just the rth order statistic gives only marginal increase in estimation accuracy. In fact, it is evident by smaller variance, that the estimate found with only the single order statistic $x_{(16)}$ is more accurate than an estimate found by knowing all of the first 14 order statistics. Note also, as expected, the variance of both estimators decreases until the right censored value approaches n, the sample size. At that point the variances separate with the point estimate from a single order statistic gaining in variance. The form of the estimator $\hat{\theta} = c_{n,r}\, X_{(r)}$ helps explain this increase in variance for the single point estimator. Clearly the variance of the estimator $V(\hat{\theta}) = V(c_{n,r}\, X_{(r)}) = c^2_{n,r}\, V(X_{(r)})$. The explanation of the decreasing-then-increasing variance can be found in the "tension" between the two terms $c^2_{n,r}$ and $V(X_{(r)})$. For the exponential distribution it can be shown that $V(X_{(r)})$ starts low and increases as r increases. On the other hand, $c^2_{n,r}$ starts out very high and then decreases. The early extreme values of $c^2_{n,r}$ eclipse the smaller values of the variance of $V(X_{(r)})$ at the smaller values of r. This makes the variance of the estimator is high. For intermediate values of r, both terms are relatively low and the variance of the estimator is also low. However, for the larger values of r, the small values of the term $c^2_{n,r}$ are eclipsed by the much higher values of $V(X_{(r)})$ and the variance of the estimator grows again.

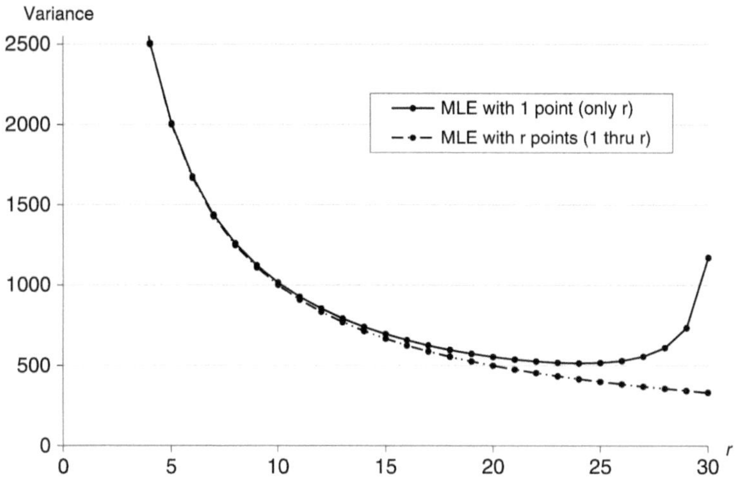

Fig. 1.1. Exact variances of the single point estimators are overlaid on the exact MLE variances for all r observations assuming the underlying distribution is exponential with $\theta = 100$ and $n = 30$

Exact interval estimates for θ can be calculated. Using the pivotal quantity method for interval estimates (see Casella and Berger [28], for example), and choosing $Q = \hat{\theta}/\theta$ as the pivotal quantity, we can find the exact distribution that is free of the parameter θ. In APPL (or by hand again using the transformation technique) one computes:

```
> Qpdf := Transform(Thetahat, [[x -> x / theta],[0, infinity]]);
```

which calculates the PDF of Q to be

$$f_Q(q) = \frac{n!\,(1 - \exp(-q/c_{n,r}))^{r-1} \exp(-(n - r + 1)\, q/c_{n,r})}{(r-1)!\,(n-r)!\,c_{n,r}} \qquad q > 0.$$

Thus, the two-tailed confidence limits for a given n and r are determined by setting

$$P\left(Q_{\alpha/2} < \frac{\hat{\theta}}{\theta} < Q_{1-\alpha/2}\right) = 1 - \alpha$$

then inverting the probability statement and isolating θ yields

$$P\left(\frac{\hat{\theta}}{Q_{1-\alpha/2}} < \theta < \frac{\hat{\theta}}{Q_{\alpha/2}}\right) = 1 - \alpha.$$

1.3 An Example for the Exponential Distribution

An illustrative example based on a previously reported problem is instructive. Meeker and Escobar [113, p. 167] describe a life test of 25 specimens of a new insulating material. The test was run until 15 items failed, with failure times 1.08, 12.20, 17.80, 19.10, 26.00, 27.90, 28.20, 32.20, 35.90, 43.50, 44.00, 45.20, 45.70, 46.30, and 47.80. Using standard MLE with Type II right censoring and the first $r = 15$ points known, the following are computed: point estimate $\hat{\theta} = 63.392$, a 95 % confidence interval for θ of $(40.48, 113.26)$, and a standard error of the estimate of $\text{SE}_{\hat{\theta}} = 16.37$. Using this paper's proposed, single order statistic estimator, the estimates are found as follows. For $n = 25$ and $r = 15$, from Table 1.1 we identify $c_{25,15} = 1.132179$, (see the table entry that is italicized). Therefore, we can derive the distribution and the estimate for the parameter using the following relationship $\hat{\theta} = 1.132179\, X_{(15)}$. For $x_{(15)} = 47.80$, we calculate the point estimate $\hat{\theta} = 1.132179 \times 47.80 = 54.1181$. Using the transformation of variables technique in APPL, the PDF of the estimator is

$$f_{\hat{\theta}}(y|\theta) = \frac{1}{\theta}\, 43307088.20\, (1 - \exp(-0.883\, y/\theta))^{14} \exp(-9.72\, y/\theta) \quad y > 0,$$

and the PDF of the pivot quantity $Q = \hat{\theta}/\theta$ is

$$f_Q(q) = 43307088.20\, (1 - \exp(-0.883\, q))^{14} \exp(-9.72\, q) \qquad q > 0.$$

From the first distribution the mean of the estimate is calculated to be $E(\hat{\theta}) = 1.004231\theta$ which is a bias of less than one-half percent larger than the true value of θ. Also, that distribution yields a standard deviation (substituting $\hat{\theta}$ in for θ) of $SE_{\hat{\theta}} = 14.493$. Finally, from the pivot point distribution, we get two critical values of the confidence interval to be $Q_{1-\alpha/2} = 1.59692$ and $Q_{\alpha/2} = 0.5534$ for a 95 % confidence interval for θ of $(33.89, 97.79)$. These compare vary favorably to the standard MLE method that required knowing the value of all 15 failure times, reported in Meeker and Escobar.

As an interesting extension, one could examine the behavior of the estimates at each of the last six single order statistics to investigate how the estimates were changing. Table 1.3 lists estimates, confidence intervals and standard errors for each of the previous order statistics. As expected, as

Table 1.3. Estimates from single order statistics that relate to the example from Section. 1.3

r	$x_{(r)}$	$\hat{\theta}$	95 % C. I.	$SE_{\hat{\theta}}$
15	47.8	54.11	$33.89 < \theta < 97.79$	14.93
14	46.3	58.37	$36.16 < \theta < 107.76$	16.07
13	45.7	64.32	$39.35 < \theta < 121.70$	18.27
12	45.2	71.27	$42.97 < \theta < 138.78$	20.98
11	44.0	78.13	$46.30 < \theta < 157.29$	23.92
10	43.5	87.56	$50.86 < \theta < 183.30$	28.03

r decreases, the standard errors of the estimates increase and the intervals get wider. Interestingly, the estimates themselves became larger as r becomes smaller. A quick verification of the standard MLEs using all 15 points showed similar behavior. This type of analysis lends itself to some extensions, such as considering linear combinations of these estimates to see if there is improvement.

1.4 The Rayleigh and Weibull Distribution Extensions

Analytic results are also attainable for the Rayleigh distribution and, to some extent, the Weibull distribution. Consider the Weibull distribution with PDF

$$f_X(x) = \theta^{-\kappa} x^{\kappa-1} \kappa\, e^{-\theta^{-\kappa} x^{\kappa}} \qquad x,\ \kappa,\ \text{and}\ \theta > 0.$$

Analytic results were not found for both parameters simultaneously. However, when $\kappa = 2$, the Weibull distribution becomes the Rayleigh distribution. Thus, the following analysis was conducted assuming a Weibull distribution with κ known. Clearly this condition has limited value for the general situation of the Weibull distribution, but it is very applicable for the Rayleigh distribution. In exploratory simulations of the Rayleigh distribution, the same decreasing variance was noted for the single point estimators. This effect was the catalyst to find analytic results for the Rayleigh distribution that follow.

Similar to what was described in Section 1.2, the likelihood function in the case in which only the rth ordered observation is known is proportional to

$$L(\theta) = F(x_{(r)})^{(r-1)} \times f(x_{(r)}) \times S(x_{(r)})^{(n-r)}.$$

Letting $x^\star = (x_{(r)}/\theta)^\kappa$ the log-likelihood function is

$$\ln L(\theta) = r \ln(1 - e^{-x^\star}) - \ln(1 - e^{-x^\star}) - \kappa \ln\theta + \kappa \ln x_{(r)} - \ln x_{(r)} + \ln\kappa$$
$$-x^\star - n\, x^\star + r\, x^\star.$$

The first partial derivative of ln $L(\theta)$ with respect to θ is

$$\frac{\partial \ln L(\theta)}{\partial \theta} = \frac{\kappa\left(-e^{x^\star} + 1 + x^\star e^{x^\star} + n x^\star e^{x^\star} - n x^\star - r x^\star e^{x^\star}\right)}{\theta\left(e^{x^\star} - 1\right)}.$$

Setting the first derivative equal to zero and solving for θ results in the following equation for the estimator

$$\hat\theta = c_{n,r}\, x_{(r)},$$

for

$$c_{n,r} = \left(\frac{1}{\text{RootOf}(\mathit{z}e^{\mathit{z}} + \mathit{z}ne^{\mathit{z}} - \mathit{z}n - \mathit{z}re^{\mathit{z}} - e^{\mathit{z}} + 1)}\right)^{1/\kappa}.$$

For the case of the Rayleigh distribution, $\kappa = 2$, it can be shown that the constants $c_{n,r}$ for the Rayleigh distribution are merely the square root of the constants for the exponential distribution with the same values of n and r, thus, the values in Table 1.1 can be appropriately adjusted to derive these constants. Likewise, for the Weibull case in which κ is known, the constants $c_{n,r}$ for the Weibull distribution are merely the constants for the exponential distribution raised to the $1/\kappa$ power.

Variances for the single order statistic and r order statistics estimators are compared for the Rayleigh distribution in Figure 1.2. The upper points are the exact variances of the single point estimators. The lower points are estimated variances (found with Monte Carlo simulation, as exact variances are not available for the Rayleigh) for the estimates from using all r order statistics. The marginal increase in accuracy from knowing all r values is again small. As in the case of the exponential distribution, distribution functions for the estimator and the pivotal quantity are calculable for the Rayleigh distribution. The PDF of the rth order statistic from the Rayleigh distribution is

$$f_{X_{(r)}}(x_{(r)}) = \frac{2nx\exp(-x_{(r)}^2/\theta^2)\left(1-\exp(-x_{(r)}^2/\theta^2)\right)^{r-1}\left(\exp(-x_{(r)}^2/\theta^2)\right)^{n-r}}{(r-1)!(n-r)!\theta^2}$$

for $x_{(r)} > 0$. Transforming that distribution by the relationship $\hat{\theta} = c_{n,r}X_{(r)}$ produces the distribution of the estimator

$$f_{\hat{\theta}}(y|\theta) = \frac{2ny\exp\left[-(n-r+1)y^2/(\theta^2c_{n,r}^2)\right]\left(1-\exp\left[-y^2/(\theta^2c_{n,r}^2)\right]\right)^{r-1}}{(r-1)!(n-r)!\theta^2c_{n,r}^2}$$

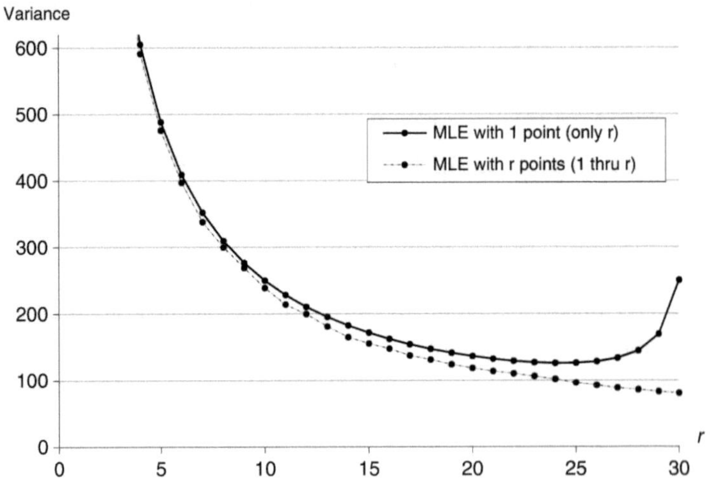

Fig. 1.2. Exact variances for the single point estimators compared to simulated variance of the r-point estimator for the Rayleigh distribution with $\theta = 100$ and $n = 30$

for $y > 0$. This distribution allows us to calculate bias and variance of the estimator, which behaves in a similar fashion to the exponential distribution. The pivotal quantity $Q = \hat{\theta}/\theta$ is again used to transform the estimator distribution into a pivotal quantity with PDF

$$f_Q(q) = \frac{2nq \exp\left[-(n-r+1)q^2/c_{n,r}^2\right]\left(1-\exp\left[-q^2/c_{n,r}^2\right]\right)^{r-1}}{(r-1)!(n-r)!c_{n,r}^2} \qquad q > 0.$$

With this distribution, confidence intervals for the Rayleigh can be formed in a similar manner as the exponential distribution. Alternatively, transformations from the Rayleigh to the exponential distributions can be performed to arrive at similar results.

1.5 Simulations and Computational Issues

The phenomenon of decreasing variance for single point order statistic estimators was noticed in some other distributions as well. For the simplistic case of the gamma distribution in which one parameter was known, but the other was not, a similar result was inferred from a simulation. In this case, the gamma distribution with PDF

$$f(x) = \frac{\alpha(\alpha\, x)^{\beta-1}e^{-\alpha\, x}}{\Gamma(\beta)} \qquad x,\ \alpha, \text{and } \beta > 0,$$

was explored. The gamma distribution simulation (as well as a similar Weibull simulation) only produced this phenomenon when the shape parameter, in this case β, was fixed. This, again, is not a desired condition, as in most two-parameter survival distributions, both parameters need to be estimated simultaneously. Furthermore, analytic results were not attainable for the gamma distribution. Figure 1.3 shows the result of estimating the one free parameter α with only one order statistic in the case of underlying gamma data for $n = 30$. Again, we see this notion that the variance of estimating with only one order statistic is almost as good as knowing all r order statistics. Note that in Figure 1.3 the simulation stopped at the $r = 28$ censoring value. This stopping point was arbitrary and could have been extended, producing a similar divergence in the two variances.

A note on computational issues is worthwhile. This analysis was done in APPL running inside of a Maple worksheet. Some of the code is included in this article for explanatory reasons. While most of this work could have been done by hand, the step of creating the $c_{n,r}$ constants required solving for θ the first derivative set equal to zero. The result of this step is an expression that needed roots, shown as `RootOf(·)` in Maple output. The function itself needs to be solved numerically, but if left alone in `RootOf` form, it represents an exact relationship. Finding this root is not easily done by hand, if at all, yet is straightforward with a working knowledge of computer algebra software,

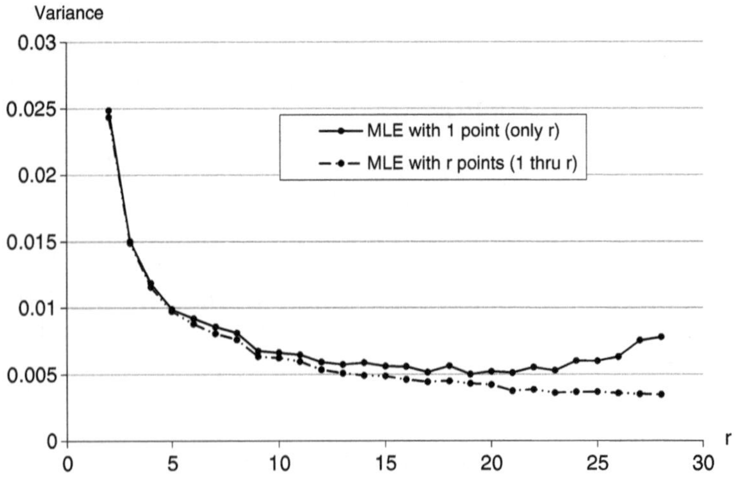

Fig. 1.3. Results of Monte Carlo simulation estimating variance of the two estimators of α for the gamma distribution with $n = 30$

especially enhanced with the procedures of APPL. In fact, there is a steadily growing number of results being published that use computer algebra software to do some or all of the analysis. A list of these types of articles is available from the author. Unlike statistical software, APPL is designed to work with symbolic expressions, not just numerical values, in order to generate new distributions. Another software package, *MathStatica* [139], is similar to APPL in that it uses procedures in Mathematica to do mathematical statistics operations. This software is not open source, however.

1.6 Implications for Design of Life Tests

These results create certain practical opportunities for improving the execution of some life tests. Although limited to only the exponential and Rayleigh distributions, this relatively simple method could be useful in input modeling for simulations. Simple input models can be accurately estimated with only one order statistic. If more than one order statistic is known, standard MLE estimates are still attainable, but with only a marginally smaller variance of the estimate, approaching that of the variance when all r lifetimes are known. For those life tests needing only a scale parameter estimate, it is possible to consider using fewer resources in observing failures. Perhaps in studies that take a long time to complete, it is be possible to eliminate continuous monitoring by evaluators and go to only periodic monitoring. Furthermore, combining these results with calculating an expected time to completion (see Leemis [93, p. 213], for example) one can design experiments with very high n, planning on terminating the experiment at the moderately low time of a smaller r

value, say $n = 50$ and $r = 7$. The result will be a shortened time on test, with reasonable variance of the estimators, should there be any left-censoring. Also, the increasing right tails in Figures 1.1, 1.2, and 1.3 do not necessarily present a problem concerning the left-censored estimation. Clearly, by the time the later order statistics occur, the experiment is so far along towards termination, that it is unreasonable in most cases to think that one could not observe that many failures over such a long time.

1.7 Conclusions

In this paper, results are presented that indicate little accuracy is lost in left-censoring on a life test. When only a single parameter is needed, even extreme left-censored data sets give almost the same quality of the estimator. Analytic results are presented for point and interval estimates for the exponential and Rayleigh distribution in the case in which only one order statistic is known. Exact distributions for the estimators are calculated, with a number of advantages that go along with knowing that distribution. Also, computational issues are presented that explain some advantages of doing mathematical statistics operations in a computer algebra system, a practice that is not very often mentioned in the literature. Clearly, more practical conditions require further research, not the least of which is considering estimating two or more parameters simultaneously. The author is also conducting similar research on distributions with left tails, such as the normal, and also the case were only a few order statistics are observed, not necessarily consecutive ones.

2

On the Inverse Gamma as a Survival Distribution

Andrew G. Glen

Abstract This paper presents properties of the inverse gamma distribution and how it can be used as a survival distribution. A result is included that shows that the inverse gamma distribution always has an upside-down bathtub (UBT) shaped hazard function, thus, adding to the limited number of available distributions with this property. A review of the utility of UBT distributions is provided as well. Probabilistic properties are presented first, followed by statistical properties to demonstrate its usefulness as a survival distribution. As the inverse gamma distribution is discussed in a limited and sporadic fashion in the literature, a summary of its properties is provided in an appendix.

Keywords Classification of distributions • Hazard functions • Transformations of random variables

---

This paper, originally published in *The Journal of Quality Technology*, Volume 43 in 2011, is another paper that relied very heavily on APPL procedures. The procedure `Transform`, which derives the PDF of transformations of random variables was key to exploring various uses for the inverse of the gamma distributions. The procedure `Product` also helped derive the newly found distribution, which the author calls the Gamma Ratio distribution, a new one-parameter lifetime distribution. Creating distributions and determining their many probabilistic properties is exactly what APPL excels at.

A.G. Glen (✉)
Department of Mathematics and Computer Science, The Colorado College, Colorado Springs, CO, USA
e-mail: andrew.glen.1984@gmail.com

A.G. Glen, L.M. Leemis (eds.), *Computational Probability Applications*, International Series in Operations Research & Management Science 247,
DOI 10.1007/978-3-319-43317-2_2

2.1 Introduction

The inverse gamma distribution with parameters α and β, $IG(\alpha, \beta)$, is mentioned infrequently in statistical literature, and usually for a specific purpose. Also called the inverted gamma distribution or the reciprocal gamma distribution, it is most often used as a conjugate prior distribution in Bayesian statistics. This article has three primary contributions regarding the IG distribution. First it proves that the IG distribution falls into the class of upside-down bathtub shaped (UBT) distributions, an unreported result. Secondly this paper demonstrates how the IG distribution can be used as a survival distribution, a use that appears unreported. Thirdly this paper collects the scattered properties of the IG distribution into one source, similar to the type of summary found in popular works like Evans et al. [50].

The number of established UBT distributions is relatively small compared to the more common increasing failure rate distributions. However, UBT distributions have been shown to provide excellent models for certain situations. For example, Aalen and Gjessing [1] show that the UBT class of distributions are good models of absorption times for stochastic models called Wiener processes. Also, Bae et al. [5] show that degradation paths often are modeled by a restricted set of distributions, with certain paths requiring UBT distributions. Crowder et al. [38] have a conjecture that UBT distributions best model a particular data set of ball bearing failure times. Lai and Xie [85] point out that UBT distributions model certain mixtures as well as failure time models in which failure is primarily due to fatigue or corrosion. To date, the well known UBT distributions are the inverse Gaussian, log-logistic, and the log-normal. Thus, a need for more UBT distributions is certainly recognized. We will show that the IG distribution fits in a moment ratio diagram in between the log-logistic and the log-normal, thus, filling a void in the UBT area (see Figure 2.3).

One primary use of the IG distribution is for Bayesian estimation of the mean of the one parameter exponential distribution (see for example Johnson et al. [72, p. 524] or Phillips and Sweeting [131, p. 777]), as well as estimating variance in a normal regression (see for example Gelman et al. [57]). It is one of the Pearson Type V distributions, as is the Wald distribution (a.k.a. the inverse Gaussian distribution, see Johnson et al. [72, p. 21]). A number of brief descriptions of the properties of the distribution are available, mostly in text books on Bayesian methods, often in the econometrics literature, e.g., Poirier [133] and Koch [81]. Kleiber and Kotz [78] list some basic properties of the IG distribution and also model incomes with the distribution. Milevsky and Posner [115] discuss the inverse gamma distribution and point out that estimation by the method of moments is tractable algebraically. There is a different distribution with, coincidentally, the same name in Zellner [175] that is derived with the square root of the inverse of a gamma random variable. Witkovsky [169, 170] derived the characteristic function of the inverse gamma. The most complete listing of some of the properties of the inverse gamma distribution is found on the Wikipedia website at `http://en.wikipedia.org/wiki/Inverse_gamma_distribution`, an anonymously written summary that does not list any sources or references. That summary is limited to the basic properties, the PDF, CDF, MGF, characteristic function, median, entropy, and the first four moments. There appears to be no comprehensive effort in the literature to employ the IG distribution as a survival distribution. This article will do that, specifically exploring the probabilistic and statistical properties of the IG distribution. Further, it is shown that the IG distribution fills a void in the available UBT distribution list, as seen in a moment ratio diagram to follow.

2.2 Probabilistic Properties

This section presents a number of probabilistic properties that are useful when considering the IG distribution as a survival distribution. An inverse gamma random variable X can be derived by transforming a random variable $Y \sim \text{gamma}(\alpha, \beta)$ with the multiplicative inverse, i.e., $X = 1/Y$. Thus, for the gamma PDF

$$f_Y(y) = \frac{\beta^\alpha}{\Gamma(\alpha)} y^{\alpha-1} e^{-\beta y} \qquad y, \alpha, \beta > 0,$$

the resulting distribution, the $IG(\alpha, \beta)$, has PDF, CDF, survivor function, and hazard function (HF)

$$\begin{aligned}
f(x) &= \frac{\beta^\alpha}{\Gamma(\alpha)} \left(\frac{1}{x}\right)^{\alpha+1} e^{-\beta/x}, \\
F(x) &= \frac{\Gamma(\alpha, \beta/x)}{\Gamma(\alpha)}, \\
S(x) &= 1 - \frac{\Gamma(\alpha, \beta/x)}{\Gamma(\alpha)}, \\
h(x) &= \frac{f(x)}{1 - F(x)} = \frac{\beta^\alpha}{\Gamma(a, 0, \beta/x)} \left(\frac{1}{x}\right)^{\alpha+1} e^{-\beta/x},
\end{aligned}$$

all with x, shape α, and scale $\beta > 0$. Recall that $\Gamma(\cdot)$ is the Euler gamma function, $\Gamma(\cdot, \cdot)$ is the incomplete gamma function, and $\Gamma(\cdot, \cdot, \cdot)$ is the generalized incomplete gamma function (see Wolfram [172] for example). Figures 2.1 and 2.2 show various PDFs and HFs for certain parameter values to indicate the shape of these functions. One of the more important aspects of a survival distribution is the shape of its hazard function. The four main classes of hazard functions for survival distributions are increasing failure rate (IFR), decreasing failure rate (DFR), bathtub-shaped (BT) and upside-down bathtub shaped (UBT). Appendix 1 shows that the IG distribution will always have a UBT hazard function.

In addition to the IG distribution always being in the UBT class, there are a number of important properties that the IG distribution has as a survival distribution.

- Moments are calculable, and the rth moment about the origin is as follows

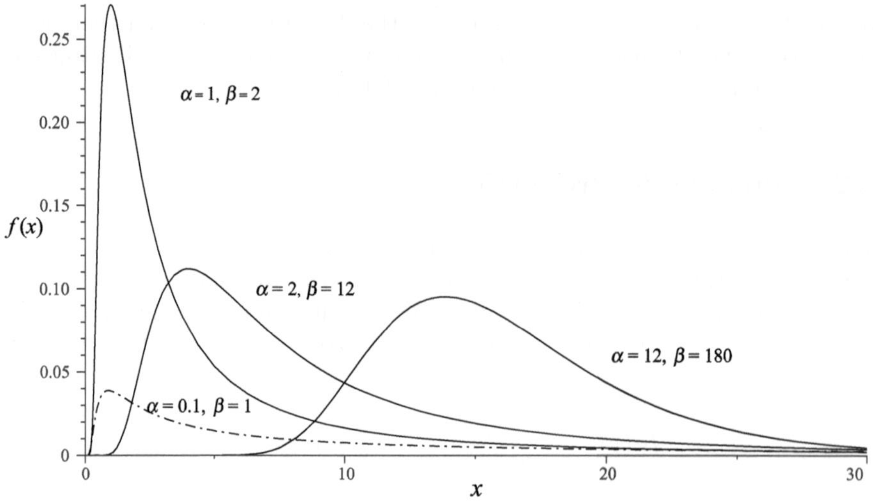

Fig. 2.1. Examples of PDFs of the inverse gamma distribution. Note the dashed PDF has a heavy right-hand tail, resulting in an undefined mean because $\alpha < 1$

$$E(X^r) = \int_0^\infty x^r f(x)dx = \frac{\beta^r \Gamma(\alpha - r)}{\Gamma(\alpha)}, \qquad \alpha > r.$$

This function is sometimes referred to as the moment function, and while it is typical that $r = 1, 2, \dots$, the function holds true for non negative real values of r. Some asymptotic results also are calculable:

$$\lim_{\alpha \to \infty} E(X^r) = 0 \qquad \text{and} \qquad \lim_{\alpha \to 0} E(X^r) = \infty.$$

- Method of moments estimation techniques are straightforward, because the mean and variance are expressed in closed-form. The first two moments about the mean are

$$\mu = \frac{\beta}{\alpha - 1}, \ \alpha > 1 \qquad \text{and} \qquad \sigma^2 = \frac{\beta^2}{(\alpha - 1)(\alpha - 2)^2}, \ \alpha > 2,$$

so the method of moments estimators can be found by algebraically inverting the set of equations in which the sample moments are set equal to the distribution moments, i.e.,

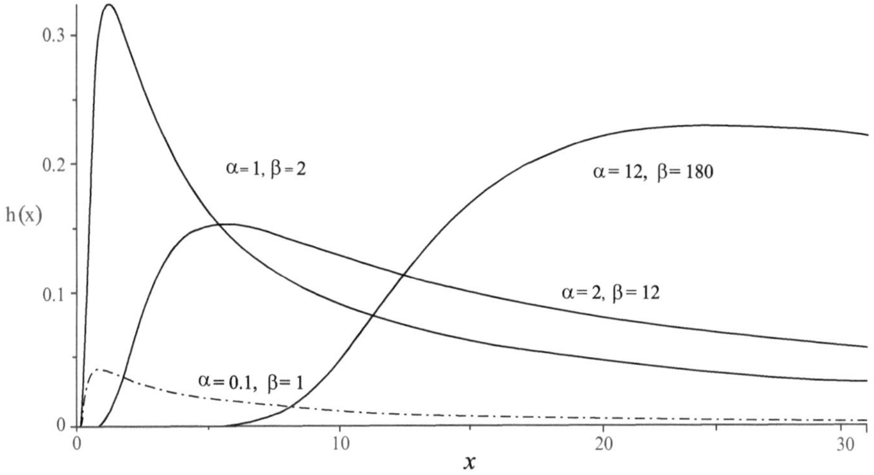

Fig. 2.2. Examples of the inverse gamma distribution hazard functions for various parameters. Note the dashed HF has a heavy tail, resulting in an undefined mean because $\alpha < 1$

$$\hat{\alpha} = \frac{\bar{x}^2 + 2s}{s} \qquad \text{and} \qquad \hat{\beta} = \frac{\bar{x}\,(\bar{x}^2 + s)}{s},$$

in which $\bar{x}$ is the sample mean and s is the sample standard deviation. This relationship is useful for finding initial estimate for numeric solutions of the MLEs.

- The limiting distribution as $\alpha \to \infty$ is degenerate at $x = 0$.
- One comparison technique when choosing among survival distributions is to plot the coefficient of variation $\gamma_2 = \sigma/\mu$ versus skewness $\gamma_3 = \dfrac{E((X-\mu)^3)}{\sigma^3}$, see Cox and Oakes [35, p. 27]. A modeler would plot $(\hat{\gamma_2}, \hat{\gamma_3})$ on such a graph to see which curve appears to be closest to the data as a start to model selection. Lan and Leemis [86] expand that graph to include the logistic-exponential distribution. Figure 2.3 takes their figure and adds the *IG* curve to the set of distributions that now include the Weibull, gamma, log-logistic, log-normal, log-exponential, inverse gamma, and the exponential distributions. The curve for the *IG* distribution falls in between the log-logistic and the log-normal distributions, in effect, helping to fill the gap between those two well-known UBT survival distributions. A more complete listing of moment ratio diagrams can be found in Vargo et al. [163], which includes the inverse gamma distribution in its figures.
- Closed-form inverse distribution functions do not exist for the *IG* distribution, so calculating quantiles, to include the median, must be done with numerical methods.
- Variate generation for the *IG* distribution can be done by inverting a variate from a gamma distribution. However, gamma variate generation

is not straightforward (as the gamma inverse distribution function (IDF) also does not exist in closed-form). Leemis and Park [96], Chap. 7, provide an explanation of various algorithms for generating gamma variates.

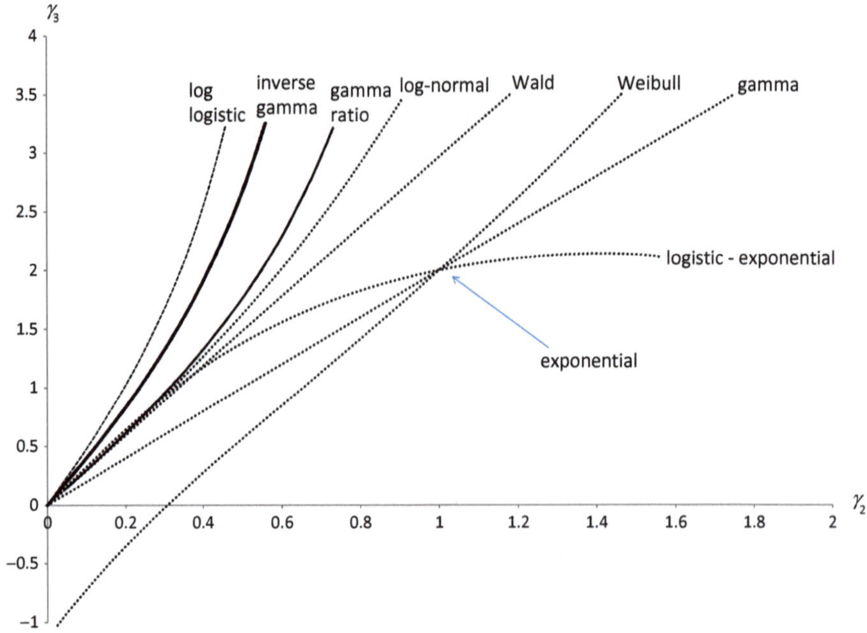

Fig. 2.3. Various two-parameter survival models with corresponding coefficient of variation versus skewness plotted. The *heavier solid line* is the inverse gamma distribution and the *lighter solid line* is the gamma ratio distribution

- Special cases and transformations of the IG distribution are equivalent to other known distributions (beside the obvious inverse transformation back to a gamma distribution). The $IG(1, \lambda)$ distribution is the inverse exponential distribution with PDF

$$f(x) = \frac{\lambda e^{-\lambda/x}}{x^2} \qquad \lambda, x > 0.$$

The $IG(\nu/2, 1/2)$ distribution is the inverse χ^2_ν distribution with PDF

$$f(x) = \frac{(\nu/2)^{\nu/2}\, x^{-1-\nu/2}\, e^{-\nu/(2x)}}{\Gamma(\nu/2)} \qquad x, \nu > 0.$$

The $IG(1/2, c/2)$ distribution is the Levy distribution with PDF

$$f(x) = \frac{(c/(2\pi))^{1/2}\, e^{-c/(2x)}}{x^{3/2}} \qquad x, c > 0.$$

The $IG(\nu/2, s^2/2)$ distribution is also called the scaled inverse χ^2 distribution, which is the form of the IG distribution that is typically used for Bayesian estimation of σ^2 in normal regression (see for example Robert [137]).

- The negative log-gamma distribution is obtained by letting $Y \sim IG(\alpha, \beta)$ and deriving the PDF of $X = \ln Y$ to be

$$f(x) = \frac{\beta^{\alpha} e^{-\alpha x - \beta e^{-x}}}{\Gamma(\alpha)} \qquad -\infty < x < \infty.$$

The log-gamma distribution is a well-behaved distribution with calculable moments and derivable inference, e.g., see Lawless [90]. Clearly, the negative log-gamma distribution is similarly behaved.

- An interesting new one-parameter survival distribution, which will be called the gamma ratio distribution, is derived as follows. Let $Y \sim$ gamma(α, β) and $X \sim IG(\alpha, \beta)$ be independent random variables. The distribution of $V = XY$ has PDF

$$f(v) = \frac{v^{\alpha-1}\Gamma\left(\alpha + \frac{1}{2}\right)\left(\frac{1}{4} + \frac{1}{2}v + \frac{1}{4}v^2\right)^{-\alpha}}{2\sqrt{\pi}\Gamma(\alpha)} \qquad \alpha, v > 0.$$

Note, this distribution is alternately formed by the ratio of two iid IG distributed random variables (which is the same as the ratio of two iid gamma distributed random variables). The rth moments about the origin for V are calculable,

$$E(V^r) = \int_0^{\infty} v^r f(v)dv = \frac{2^{2\alpha-1}\sqrt{2}\Gamma(\alpha - r)\Gamma(\alpha + r)}{2^{2\alpha-\frac{1}{2}}\left(\Gamma(\alpha)\right)^2} \qquad \alpha > r$$

and the mean is close to one, as is expected by its construction,

$$\mu_V = \frac{\alpha}{\alpha - 1} \qquad \alpha > 1$$

with variance

$$\sigma_V^2 = \frac{(2\alpha - 1)\alpha}{(\alpha - 2)(\alpha - 1)^2} \qquad \alpha > 2.$$

The CDF and hazard function are calculable, but require special functions in Maple. The gamma ratio distribution is of interest because it joins the exponential and the Rayleigh distributions as a one parameter survival distribution. For parameter values of $\alpha > 1$ it can be shown to have a UBT failure rate, however for $0 < \alpha \leq 1$ it appears to have a decreasing failure rate. This conjecture still needs to be proven, but can be shown anecdotally. When $\alpha \leq 1$ the distribution has a very heavy right tail, further indicating that no first moment exists. Furthermore, the PDF is hump-shaped for $\alpha > 1$, as is the Rayleigh, a shape the exponential can not attain. The gamma ratio distribution fits nicely in the moment ratio diagram, see Figure 2.3, as it further fills the gap between the log-logistic and the log-normal distributions. A disadvantage to this distribution being used as a survival distribution is that the parameter α is a shape parameter, but not a scale parameter, thus, limiting its flexibility as units of measure change. This distribution warrants further research as another survival distribution, as it is in a small class of one-parameter distributions as well as in the UBT class.

- The distribution of a sum of iid IG random variables does not produce a closed-form PDF. However, products are better behaved. Let $X_i \sim IG(\alpha, \beta)$, for $i = 1, 2$ be iid, then $Z = X_1 X_2$ has PDF

$$f(z) = \frac{2z^{-1-\alpha}\beta^{2\alpha}\texttt{BesselK}\left(0, \frac{2\beta}{\sqrt{z}}\right)}{\left(\Gamma(\alpha)\right)^2} \qquad z > 0.$$

2.3 Statistical Inference

In order for a survival distribution to be viable for empirical modeling, statistical methods must be reasonably tractable. The inverse gamma distribution can produce statistical inference for both complete and right-censored data sets. Some likelihood theory and examples of fitting each type of data set are presented.

2.3.1 Complete Data Sets

For the uncensored case, let t_1, t_2, ..., t_n be the failure times from an experiment. The likelihood function is

$$L(\alpha, \beta) = \prod_{i=1}^{n} f(t_i, \alpha, \beta) = \prod_{i=1}^{n} \frac{\beta^\alpha}{\Gamma(\alpha)} \left(\frac{1}{t_i}\right)^{\alpha+1} e^{-\beta/t_i}.$$

Taking the natural logarithm and simplifying produces

$$\ln L(\alpha, \beta) = n\,\alpha \ln\beta - n\ln\bigl(\Gamma(\alpha)\bigr) + (\alpha+1)\sum_{i=1}^{n} \ln\left(\frac{1}{t_i}\right) - \beta\sum_{i=1}^{n}\frac{1}{t_i}.$$

The first partial derivatives of $\ln L(\alpha, \beta)$ with respect to the two parameters are

$$\frac{\partial \ln L(\alpha, \beta)}{\partial \alpha} = n\ln\beta - n\Psi(\alpha) + \sum_{i=1}^{n} \ln\left(\frac{1}{t_i}\right)$$

and

$$\frac{\partial \ln L(\alpha, \beta)}{\partial \beta} = \frac{n\alpha}{\beta} - \sum_{i=1}^{n}\frac{1}{t_i},$$

for $\Psi(\alpha) = \frac{d}{d\alpha} \ln \Gamma(\alpha)$ is the digamma function. Equating these two partial derivatives to zero and solving for the parameters does not yield closed-form solutions for the maximum likelihood estimators $\hat{\alpha}$ and $\hat{\beta}$ but the system of equations is well behaved in numerical methods. If initial estimates are needed, the method of moments can be used. When the equations are set equal to zero, one finds $\beta = n\alpha(\sum t_i^{-1})^{-1}$, which reduces the problem to a single parameter equation

$$n \ln n + n \ln \alpha - \ln \left(\sum_{i=1}^{n} t_i^{-1} \right) - n\Psi(\alpha) - \sum_{i=1}^{n} \ln t_i = 0,$$

which must be solved by iterative methods.

Confidence intervals for the MLEs can be obtained with the observed information matrix, $O(\hat{\alpha}, \hat{\beta})$. Cox and Oakes [35] show that it is a consistent estimator of the Fisher information matrix. Taking the observed information matrix

$$O(\hat{\alpha}, \hat{\beta}) = \left(\begin{array}{cc} \frac{-\partial^2 \ln L(\alpha,\beta)}{\partial \alpha^2} & \frac{-\partial^2 \ln L(\alpha,\beta)}{\partial \alpha \partial \beta} \\ \frac{-\partial^2 \ln L(\alpha,\beta)}{\partial \beta \partial \alpha} & \frac{-\partial^2 \ln L(\alpha,\beta)}{\partial \beta^2} \end{array} \right)_{\alpha=\hat{\alpha}, \beta=\hat{\beta}},$$

one then inverts the matrix and uses the square root of the diagonal elements as estimates of the standard deviations of the MLEs to form confidence intervals.

To illustrate the use of the inverse gamma distribution as a survival distribution, consider Lieblein and Zelen's [101] data set of $n = 23$ ball bearing failure times (each measurement in 10^6 revolutions):

17.88	28.92	33.00	41.52	42.12	45.60	48.48
51.84	51.96	54.12	55.56	67.80	68.64	68.64
68.88	84.12	93.12	98.64	105.12	105.84	127.92
128.04	173.40					

This is an appropriate example because Crowder et al. [38, p. 63] conjectured that UBT shaped distributions might fit the ball bearing data better than IFR distributions based on the values of the log likelihood function at the maximum likelihood estimators. Using Maple's numeric solver `fsolve()`, the MLEs are $\hat{\alpha} = 3.6785, \hat{\beta} = 202.5369$. Figure 2.4 gives a graphical comparison of the survivor functions for the Weibull and inverse gamma distributions fit to the empirical data.

The observed information matrix can be calculated from the MLEs of the ball bearing set:

$$O(\hat{\alpha}, \hat{\beta}) = \begin{pmatrix} 7.1783 & -0.1136 \\ -0.1136 & 0.0021 \end{pmatrix}.$$

The inverse of this matrix gives the estimate of the variance–covariance matrix for the MLEs:

$$O^{-1}(\hat{\alpha}, \hat{\beta}) = \begin{pmatrix} 1.080 & 59.472 \\ 59.472 & 3759.361 \end{pmatrix}.$$

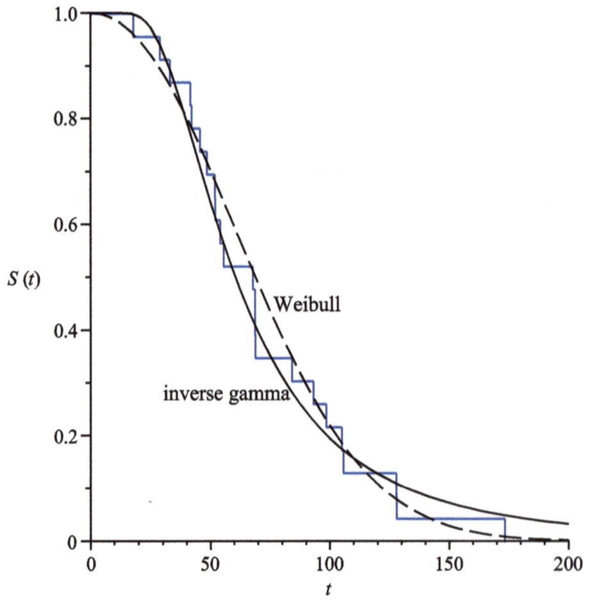

Fig. 2.4. The empirical, fitted inverse gamma, and fitted Weibull distributions for the ball bearing data set

The square roots of the diagonal elements give estimates of standard deviations for the MLEs $\text{SE}_{\hat{\alpha}} = \sqrt{1.080} \cong 1.039$ and $\text{SE}_{\hat{\beta}} = \sqrt{3759} \cong 61.31$. Thus, the approximate 95 % confidence intervals for the estimates are

$$1.550 < \alpha < 5.807 \qquad \text{and} \qquad 85.511 < \beta < 319.563.$$

The off-diagonal elements give us the covariance estimates. Note the positive covariance between the two estimates, a fact that is made evident in the following method for joint confidence regions.

There are a number of different methods to get joint confidence regions for these two estimates. Chapter 8 of Meeker and Escobar [113] gives a good summary of many of these techniques. One such method, finding the joint confidence region for α and β, relies on the fact that the likelihood ratio statistic, $2\left(\ln L(\hat{\alpha}, \hat{\beta}) - \ln L(\alpha, \beta)\right)$, is asymptotically distributed as a χ^2_2 random variable. Because $\chi^2_{2,0.05} = 5.99$, the 95 % confidence region is the set of all (α, β) pairs satisfying

$$2(\ln L(\hat{\alpha}, \hat{\beta}) - \ln L(\alpha, \beta)) = 2(-114.154 - \ln L(\alpha, \beta)) < 5.99.$$

The boundary of the region, found by solving this inequality numerically as an equality, is displayed in Figure 2.5 for the ball bearing failure times. Note in this figure the positive correlation between the two parameter estimates, displayed by the positive incline to the shape of the confidence region.

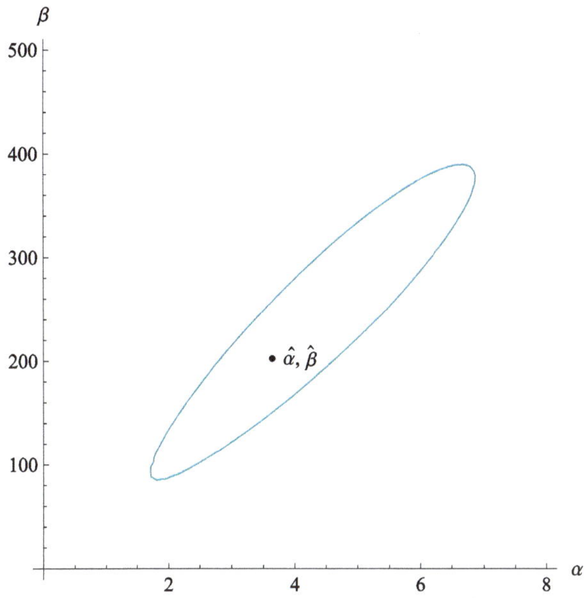

Fig. 2.5. The 95 % joint confidence region for α and β for the ball bearing data

Comparisons can be made with other survival distributions. Lan and Leemis [86], as well as Glen and Leemis [61], compare the Kolmogorov–Smirnov (K–S) goodness of fit statistic D_{23} at their MLE values for a number of typical survival distributions. Table 2.1 inserts the inverse gamma into that comparison. Note that the inverse gamma has a better fit than any of the IFR class of distributions (to include its 'parent,' the gamma distribution). The inverse gamma also fits similarly to the other UBT distributions, giving more credence to Crowder's conjecture that ball bearing data is better fit by UBT models.

2.3.2 Censored Data Sets

It is important that survival distributions are capable of producing inference for censored data as well. The statistical methods are similar to uncensored data, but the derivatives of the likelihood equation are not in closed-form, thus, the numerical methods require some more assistance in the form of initial estimates. Mirroring the process of Glen and Leemis [61] and Lan and Leemis [86], initial estimates will be derived from a "method of fractiles" estimation. The data set to be used comes from Gehan's [56] test data of remission times for leukemia patients given the drug 6-MP of which $r = 9$ observed remissions were combined with 12 randomly right censored patients. Denoting the right censored patients with an asterisk, the remission times in weeks are:

6	6	6	6*	7	9*	10
10*	11*	13	16	17*	19*	20*
22	23	25*	32*	32*	34*	35*

Table 2.1. Kolmogorov–Smirnov goodness-of-fit Statistics for the ball bearing data at the respective MLEs

Distribution	Class	D_{23}
Exponential	IFR	0.301
Weibull	IFR	0.152
Gamma	IFR	0.123
Logistic-exponential	UBT	0.109
Inverse gamma	**UBT**	**0.104**
Inverse Gaussian	UBT	0.099
Log-logistic	UBT	0.094
Arctangent	UBT	0.093
Log-normal	UBT	0.090

To fit the inverse gamma distribution to this data, let $t_1, t_2, \ldots, t_n$ be the n remission times and let $c_1, c_2, \ldots, c_m$ be the m associated censoring times. Our maximum likelihood function is derived as follows:

$$L(\alpha, \beta) = \prod_{i \in U} f(t_i, \alpha, \beta) \prod_{i \in C} S(c_i, \alpha, \beta)$$

in which U and C are the sets of indices of uncensored and censored observations, respectively. The log likelihood function does not simplify nicely as natural logarithms of $S(x)$ are functions that cannot be expressed in closed-form. Programming environments like Maple and Mathematica will produce the log likelihood function, but it is not compact, thus, it is not shown here. It is, however, available from the author. A method of fractiles initial estimate sets the empirical fractiles equal to the $S(x)$ evaluated at the appropriate observation in a 2×2 set of equations. In the case of the 6-MP data set, two reasonable choices for the equations are

$$S(10) \cong 0.80 = 1 - \frac{\Gamma(\alpha, \frac{\beta}{10})}{\Gamma(\alpha)}$$

and

$$S(21) \cong 0.55 = 1 - \frac{\Gamma(\alpha, \frac{\beta}{21})}{\Gamma(\alpha)},$$

which produce initial estimates $\hat{\alpha}_0 = 0.9033$ and $\hat{\beta}_0 = 14.6507$. Initial values of S (i.e., in this case $S(10)$ and $S(21)$) should be chosen to adequately represent a reasonable spread of the data, not being too close together, and not being too close to an edge. The goal is to find two values that will act as initial estimates for the numeric method for finding the true MLEs that will cause the numerical method to converge. One may have to iterate on finding two productive S values. Then one must use these initial estimates required by numerical methods, take the two partial derivatives of the log likelihood with respect to α and β, set them equal to zero, and solve numerically for the MLEs, which are $\hat{\alpha} = 0.9314$ and $\hat{\beta} = 15.4766$. A plot of the inverse gamma and the arctangent distributions (see Glen and Leemis [61]) along with the Kaplan–Meier non parametric estimator are presented in Figure 2.6. The inverse gamma has a superior fit in the earlier part of the distribution.

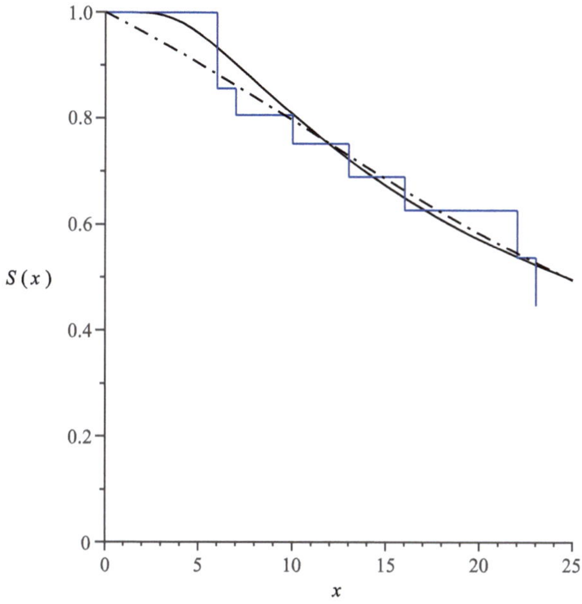

Fig. 2.6. The MLE fits of the inverse gamma distribution (*solid curve*) and the arctangent distribution (*dashed curve*) against the censored 6-MP data set

2.4 Conclusions

This paper presents the inverse gamma distribution as a survival distribution. A result is proved that shows that the inverse gamma distribution is another survival distribution in the UBT class. The well-known UBT distributions are relatively few, so it is helpful to have an alternative when dealing with UBT models. Since sporadic mention is made in the literature of the distribution, as well as very little probability or statistical information, this paper helps fill this gap in the literature. Probabilistic properties and statistical methods are provided to assist the practitioner with using the distribution. Finally, many properties, previous published in a scattered manner, are combined into this one article.

Appendix 1

In this appendix it is shown that the hazard function of the inverse gamma distribution can only have the UBT shape and the mode, $x^\star$, of the hazard function is bounded by $0 < x^\star < \frac{2\beta}{\alpha+1}$.

In order to show that the hazard function is always UBT in shape, the following is sufficient: (1) $\lim_{x\to 0} h(x) = 0$, (2) $\lim_{x\to\infty} h(x) = 0$, and (3) $h(x)$ is unimodal. Because all hazard functions are positive and have infinite area underneath the hazard curve, Rolle's theorem, taken with (1) and (2) guarantees at least one maximum point (mode). So it must be shown that there is only one maximum value of $h(x)$ on the interval $0 < x < \infty$ to prove than h is UBT in shape.

First consider (1) that $\lim_{x\to 0} h(x) = 0$. If we make a change of variable $z = 1/x$ then the limit to be evaluated becomes

$$\lim_{z\to\infty} \frac{\beta^\alpha e^{-\beta z}\, z^{\alpha+1}}{\Gamma(\alpha, 0, \beta z)}.$$

The denominator is an incomplete gamma function which in the limit becomes the complete gamma function, $\Gamma(\alpha) = \Gamma(\alpha, 0, \infty)$, a constant. The numerator can be rewritten as

$$\frac{\beta^\alpha z^{\alpha+1}}{e^{z\beta}}$$

so that it has the form ∞/∞ in the limit and can therefore be evaluated with $\lceil \alpha \rceil$ (the next highest integer if α is not an integer) successive applications of L'Hospital's rule. Thus, the limit of the numerator of h is $\lim_{z\to\infty} \beta^\alpha e^{-z\beta}\, z^{\alpha+1} = 0$. Consequently $\lim_{x\to 0} h(x) = 0$ and (1) is satisfied.

Next, consider (2) that $\lim_{x\to\infty} h(x) = 0$. Again applying L'Hospital's rule, this time only once, the first derivative of the numerator with respect to x is

$$-\beta^{\alpha} e^{-\beta/x} \left(\frac{1}{x}\right)^{3+\alpha} (-\beta + x + \alpha x).$$

The first derivative of the denominator with respect to x is

$$-\frac{\beta^{\alpha} e^{-\beta/x} \left(\frac{1}{x}\right)^{-1+\alpha}}{x^2}.$$

Dividing the numerator's derivative by the denominator's derivative and simplifying we get

$$\lim_{x \to \infty} \frac{-\beta + x + \alpha x}{x^2} = 0$$

and (2) is satisfied.

Rolle's Theorem allows us to conclude that because the left and right limits of $h(x)$ are both zero, and because h is a positive function, there must be some value ξ such that $h'(\xi) = 0$ on the interval $0 < \xi < \infty$. Because h has a left limit at the origin, it cannot have a DFR or a BT shape. Because h has a right limit at zero, it cannot have an IFR shape. Also, because h is a positive function, there is at least one critical point at $x = \xi$ that must be a maximum.

Finally, considering element (3) of the proof, one must rely on the Lemma E.1 and Theorem E.2 of Marshall and Olkin [107, pp. 134–135]. That theorem and lemma establish that

$$\rho(x) = -\frac{f'(x)}{f(x)}$$

and $h(x)$ have the same number of sign changes and in the same order. Further the critical point of ρ is an upper bound for the critical point of h. For this distribution,

$$\rho(x) = \frac{-\beta + x + \alpha x}{x^2},$$

which has first derivative

$$\rho'(x) = \frac{2\beta - (1 + \alpha)x}{x^3},$$

and can be shown to be positive from $0 < x < \frac{2\beta}{\alpha+1}$ and negative from $\frac{2\beta}{\alpha+1} < x < \infty$. Thus, h has only one change in the sign of its slope, is therefore UBT in shape, and has a single mode $x^{\star} < \frac{2\beta}{\alpha+1}$.

Appendix 2

This appendix summarizes the key properties of the inverse gamma distribution that have not already been reported in the rest of the paper. The purpose of this appendix is ensure that all the major properties of the IG distribution are in one document, as to date no manuscript appears to be this comprehensive. Most of the properties reported in the appendix are found in the works previously cited or are derived with straightforward methods.

- Mode: $\dfrac{\beta}{\alpha+1}$
- Coefficient of variation: $\sigma/\mu = (\alpha-1)^{-1/2}$
- Skewness: $E\left[\left(\dfrac{X-\mu}{\sigma}\right)^3\right] = \dfrac{4\sqrt{\alpha-1}}{\alpha-3}$ for all $\alpha > 3$
- Excess kurtosis: $E\left[\left(\dfrac{X-\mu}{\sigma}\right)^4\right] = \dfrac{30\alpha-66}{(\alpha-3)(\alpha-4)}$ for all $\alpha > 4$
- Entropy: $\alpha - (1+\alpha)\Psi(\alpha) + \ln\big(\beta\Gamma(\alpha)\big)$
- Moment generating function: $\dfrac{-2\beta t^{\alpha/2}\text{BesselK}\left(\alpha, 2\sqrt{-\beta t}\right)}{\Gamma(\alpha)}$
- Characteristic function: $\dfrac{-2i\beta t^{\alpha/2}\text{BesselK}\left(\alpha, 2\sqrt{-i\beta t}\right)}{\Gamma(\alpha)}$
- Laplace transform: $\dfrac{2\beta s^{\alpha/2}\text{BesselK}\left(\alpha, 2\sqrt{-\beta s}\right)}{\Gamma(\alpha)}$
- Mellin transform: $\dfrac{\beta^{s-1}\Gamma(1-s+\alpha)}{\Gamma(\alpha)}$

3

Order Statistics in Goodness-of-Fit Testing

Andrew G. Glen, Donald R. Barr, and Lawrence M. Leemis

Abstract A new method is presented for using order statistics to judge the fit of a distribution to data. A test statistic based on quantiles of order statistics compares favorably with the Kolmogorov–Smirnov and Anderson–Darling test statistics. The performance of the new goodness-of-fit test statistic is examined with simulation experiments. For certain hypothesis tests, the test statistic is more powerful than the Kolmogorov–Smirnov and Anderson–Darling test statistics. The new test statistic is calculated using a computer algebra system because of the need to compute exact distributions of order statistics.

Keywords Computational algebra system • Goodness-of-fit • Model adequacy • Order statistics • Power

---

Originally published in *IEEE Transactions on Reliability*, Volume 50, Number 2 in 2001, this paper is the first result using APPL to improve goodness of fit testing. Because many tests rely on the probability integral transform which makes many tests reduce to uniformity tests, APPL's `Transform` procedure, used with `OrderStat`, produces exact distributions of random PDFs of order statistics. This means that new distributions are created during the calculation of the actual test statistic. This type of creation of new distributions is not possible in standard statistical packages.

A.G. Glen (✉)
Department of Mathematics and Computer Science, The Colorado College, Colorado Springs, CO, USA
e-mail: andrew.glen.1984@gmail.com

D.R. Barr
USMA, West Point, NY, USA

L.M. Leemis
The College of William and Mary, Williamsburg, VA, USA
e-mail: leemis@math.wm.edu

A.G. Glen, L.M. Leemis (eds.), *Computational Probability Applications*, International Series in Operations Research & Management Science 247, DOI 10.1007/978-3-319-43317-2_3

Acronyms

CDF	Cumulative distribution function
iid	s-Independent and identically distributed
PDF	Probability density function
K–S	Kolmogorov–Smirnov
A–D	Anderson–Darling
MLE	Maximum likelihood estimator
APPL	A Probability Programming Language

Notation

n	Sample size
$F(x)$	CDF, $F(x) = \Pr\{X \leq x\}$
$f(x)$	PDF
$F_n(x)$	Empirical CDF
D_n	K–S test statistic
$U(0,1)$	Uniform distribution between 0 and 1
$N(\mu, \sigma^2)$	Normal distribution with mean μ and variance σ^2
$X_1, X_2, \ldots, X_n$	Random sample
$X_{(1)}, X_{(2)}, \ldots, X_{(n)}$	Order statistics
$\mathcal{P}$	$\mathcal{P}$-vector
$p_i = F_{X_{(i)}}(x_{(i)})$	ith Element of the $\mathcal{P}$-vector
$\hat{\lambda}$, $\hat{\kappa}$	Weibull distribution MLEs
θ_0	Value of the parameter θ under the null hypothesis
$1 - \beta$	Power of a hypothesis test

3.1 Introduction

There are a wide variety of applications of order statistics in statistical-inference procedures concerning:

- behavior in the tails of a parent distribution, e.g., outlier tests in Tietjen [160],
- situations in which the ordered data are collected over time, e.g., life testing in Barlow and Proschan [7],
- estimation of parameters in terms of linear combinations of order statistics, e.g., David [41],
- adapting procedures for use with censored samples, e.g., Cox and Oakes [35] and David [41].

A large literature on the subject exists (see, for example, David and Nagaraja [43] and references cited therein).

One important class of problems involving order statistics is goodness-of-fit testing see, for example D'Agostino and Stevens [40]. The Shapiro–Wilks normality test [73] is based on a statistic that uses, in part, the s-expected value of each order statistic. Because many goodness-of-fit test statistics commonly used are defined through the empirical CDF F_n (a piecewise constant function that takes an upward step of $1/n$ at each data value), they may be considered to be functions of the underlying order statistics. It is common practice to transform order statistics from a parent population for the continuous random variable X with CDF F to corresponding $U(0,1)$ order statistics through the probability integral transformation, $U = F(X)$. If the proposed test statistic is invariant under this transformation, it follows that its distribution can be derived for a $U(0,1)$ parent distribution, and applied for any other continuous parent distribution. That is, the procedure in this case is distribution-free.

An example of such a test is the K–S one-sample test, based on the test statistic $D_n = \sup_x |F(x) - F_n(x)|$. Because the L_∞ norm is invariant under transformations on x, it follows that one can find critical values for D_n using the distribution of $\max_i\{\max\{|u_i - \frac{i-1}{n}|, |u_i - \frac{i}{n}|\}\}$, for $u_i = F(x_{(i)}), i = 1, 2, \ldots, n$. Following the development of the K–S test, many articles adapted the test to related order statistic-based tests:

- tests for data on a circle, e.g., Kuiper [83] and Stephens [151],
- tests with parameters estimated from data, e.g., Lilliefors [102, 103]
- tests for censored samples, e.g., Barr and Davison [8].

In addition to its use in non-parametric goodness-of-fit testing, the K–S procedure may be used in a general way to obtain confidence regions for parameter vectors as in Barr and Zehna [9], which is a parametric problem.

Section 3.2 defines a vector of statistics obtained by transforming the order statistics through each of the n hypothesized order statistic CDFs. It suggests possible uses of the vector in a variety of statistical problems. Section 3.3 considers the practical aspects associated with determining the values in the vector. Section 3.4 applies these ideas to define a goodness-of-fit statistic. Section 3.5 uses Monte Carlo simulation to show that it's performance is comparable to the K–S and A–D tests in certain settings. Section 3.6 outlines possible extensions.

3.2 $\mathcal{P}$-Vector

The real-valued $\mathcal{P}$-vector is proposed as a measure of the goodness-of-fit of empirical data by a hypothesized distribution. Let $X_1, X_2, \ldots, X_n$ denote a random sample from a population with continuous CDF F_X, and let $X_{(1)}, X_{(2)}, \ldots, X_{(n)}$ be the corresponding order statistics. Then the individual order statistics have marginal pdfs given by [87]

$$f_{X_{(i)}}(x) = \frac{n!}{(i-1)!(n-i)!}[F_X(x)]^{i-1}[1 - F_X(x)]^{n-i} f_X(x) \qquad i = 1, 2, \ldots, n,$$

for $f_X(x) = F'_X(x)$. In principle, one can use these distributions to determine the quantiles for each ordered observation in its respective distribution. Let

$$\mathcal{P} \equiv [F_{X_{(1)}}(x_{(1)}), F_{X_{(2)}}(x_{(2)}), \ldots, F_{X_{(n)}}(x_{(n)})].$$

For notational simplicity, let $p_i = F_{X_{(i)}}(x_{(i)}), i = 1, 2, \ldots, n$. Intuitively, poor fit is indicated by extreme components in $\mathcal{P}$. Thus, under the null hypothesis H_0: X has CDF $F_X(x)$ with $n = 3$ observations, for example, a $\mathcal{P}$-vector of $[0.453, 0.267, 0.623]$ intuitively indicates a good fit more so than a $\mathcal{P}$-vector of $[0.001, 0.005, 0.997]$.

Because the p_i values are the result of the probability integral transformation, then $P_i \sim U(0,1), i = 1, 2, \ldots, n$, for any continuous population CDF F_X. While the P_i are identically distributed, they are not s-independent. There is positive autocorrelation among elements of the $\mathcal{P}$-vector.

The $\mathcal{P}$-vector has several potential uses as a basis for distribution-free statistical procedures. It can provide an distribution-free tool to identify outliers. Tietjen [160] mentions in his paper on outliers "We shall discuss here only the underlying assumption of normality because there is very little theory for any other case." As explained in Section 3.3, it is possible and practical to calculate all p_i for nearly any hypothesized continuous distribution, so one may use the $\mathcal{P}$-vector as a basis to identify outliers from any specified continuous distribution. A reasonable approach is to examine the first and last few elements of $\mathcal{P}$ (e.g., p_1 and p_n) to determine whether they are s-significant.

Example 3.1. Consider the following example, using Lieblein and Zelen's [101] data set of $n = 23$ ball bearing failure times (measured in 10^6 revolutions):

17.88	28.92	33.00	41.52	42.12	45.60	48.48	51.84
51.96	54.12	55.56	67.80	68.64	68.64	68.88	84.12
93.12	98.64	105.12	105.84	127.92	128.04	173.40	

This data set is often fitted by the Weibull distribution with MLE's $\hat{\lambda} \cong 0.0122$ and $\hat{\kappa} \cong 2.10$ [93]. The corresponding $\mathcal{P}$-vector for this set, assuming H_0 corresponds to the fitted Weibull as the underlying population distribution, is:

$$\begin{aligned}[&0.609, 0.717, 0.631, 0.755, 0.589, 0.546, 0.487, 0.456,\\ &0.296, 0.234, 0.161, 0.460, 0.334, 0.198, 0.107, 0.425,\\ &0.570, 0.569, 0.578, 0.386, 0.736, 0.460, 0.833].\end{aligned}$$

A possible interpretation of this $\mathcal{P}$-vector is that neither $x_{(1)}$ nor $x_{(23)}$ is an outlier. Furthermore, the value that is furthest from the median of its distribution is the 15th ordered data point, 68.88, with p-value equal to 0.10575. The inference here is not exact due to the use of MLEs. It is reasonable, however, to conclude that the Weibull distribution adequately models the entire range of the data set.

Other possible uses of the $\mathcal{P}$-vector are in statistical inferences involving censored sampling and estimation based on order-agreement. The main application investigated here, however, is goodness-of-fit testing using test statistics based on the $\mathcal{P}$-vector.

3.3 Computation of the $\mathcal{P}$-Vector

Computing the elements of the $\mathcal{P}$-vector can be accomplished using computational algebra languages, such as Maple [65]. APPL determines the distributions of transformations, sums, products, and order statistic distributions of continuous random variables [60]. The APPL procedure `OrderStat(X, n, r)`, for example, determines the distribution of the rth out of n order statistics. Combined with the procedure `CDF(X, x)`, which returns $F_X(x)$, the elements p_i are calculated. In the previous example, for instance, the $\mathcal{P}$-vector can be calculated and stored in the Maple list `p` with the following APPL statements:

```
> data := [17.88, 28.92, 33.00, ..., 173.40];
> n := 23;
> hat := MLEWeibull(data);
> p := [ ];
> X := WeibullRV(hat[1], hat[2]);
> for r from 1 to n do
>   Y := OrderStat(X, n, r);
>   p := [op(p), CDF(Y, data[r])];
> od;
```

It is preferable to calculate the $\mathcal{P}$-vector by first transforming to $U(0,1)$ order statistics, and then determining the quantiles p_i using corresponding beta CDFs, which are relatively simple polynomials, as indicated in Figure 3.1. The calculation of $p_i = F_{X_{(i)}}(x_{(i)})$ is depicted by the path shown by a solid line in Figure 3.1. This method of computation relies on the ability to calculate quantiles of all of the order statistics $X_{(i)}$, although recurrence relations for CDFs of order statistics might speed computation [43]. An alternate approach, using beta CDFs, is depicted by the dashed line in Figure 3.1. It requires the transformation of the $x_{(i)}$ into $U(0,1)$ random variables and then determines their quantiles using appropriate beta CDFs. A simple theorem (omitted for brevity) shows that both paths are equivalent, i.e., $F_{Z_{(i)}}(z_{(i)}) = F_{X_{(i)}}(x_{(i)})$, for $Z = F_X(X)$.

The quantile p_i following the lower dotted path in Figure 3.1 could also be considered. The path indicated in Figure 3.1 by the dashed line is generally preferred, because the distributions leading to the p_i elements are polynomials. The computations needed for the solid path are calculable, although they typically take appreciably longer to calculate. Also, the CDF of X is typically more tractable than the CDFs of the $X_{(i)}$ values.

3.4 Goodness-of-Fit Testing

In general, goodness-of-fit testing involves stating an assumed null hypothesis, H_0: $f_X(x;\theta) = f_0(x;\theta_0)$ and then assessing whether there is sufficient evidence in a random sample to reject H_0. Because the $\mathcal{P}$-vector was derived under H_0, its elements provide a direct judgment of fit. Any p_i too low or too high may indicate a poor fit. We want to form a single statistic based on the p_i's that indicate good or bad fit. Some obvious candidate test statistics are

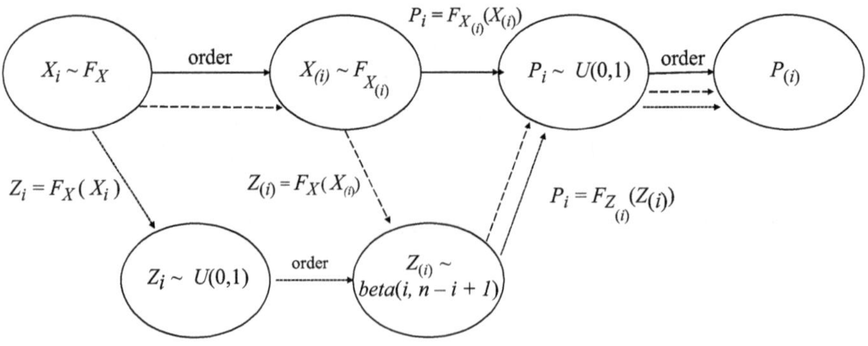

Fig. 3.1. Transformations from iid observations $X_1, X_2, \ldots, X_n$ to the sorted $\mathcal{P}$-vector elements $P_{(1)}, P_{(2)}, \ldots, P_{(n)}$

$$\sum_{i=1}^{n} |p_i - 0.5|, \; \sum_{i=1}^{n} (p_i - 0.5)^2, \; \text{and} \; \sum_{i=1}^{n} \left| p_{(i)} - \frac{i}{n+1} \right|.$$

Based on a modest set of simulations, all three of these test statistics appear to suffer from low power. Instead, we rely on a variation of the form of the A–D A^2 statistic, when calculated with the probability integral transformation [153]. The A–D A^2 statistic is developed with the $Z_{(i)}$ values which are depicted in the lower-right of Figure 3.1. We define a test statistic in terms of a linear combination of the natural logarithms of p_i and $1-p_i$. This test statistic is large whenever at least one p_i is too close to 0 or 1. Sort the elements of the $\mathcal{P}$-vector so that $p_{(1)}$ is the smallest of the elements of the $\mathcal{P}$-vector and $p_{(n)}$ is the largest. The test statistic P_s is

$$P_s \equiv -n - \frac{1}{n} \sum_{i=1}^{n} \left[(2n + 1 - 2i) \ln(p_{(i)}) + (2i - 1) \ln(1 - p_{(i)}) \right].$$

The coefficients $2n + 1 - 2i$ and $2i - 1$ are exchanged from the position they would appear in the A–D statistic. This switch resulted in higher power in our simulations, because it results in a higher test statistic for more extreme observations. We examined the power of P_s by simulation. The goodness-of-fit tests based on P_s have power about that of A^2 in most cases, and both P_s and A^2 generally out-perform the K–S test, as stated in D'Agostino and Stevens [40], for the A–D test. There is at least one case in which P_s out-performs the A–D test: the case of guarding against an improper variance parameter under the null and alternate hypotheses of normality, as illustrated in the next section.

3.5 Power Estimates for Test Statistics

The hypothesis test that uses P_s as a test statistic is appreciably more powerful than both the K–S and A–D tests in the simulation experiment described here. We approximated the power of the three tests, and a fourth test $P_s^\star$ using p_i in place of $p_{(i)}$ in the definition of P_s. The hypothesis test is

- H_0: $N(0, 1)$,
- H_1: $N(0, \sigma^2)$ for $\sigma^2 \neq 1$.

Random samples of size 10 were generated from $N(0, \sigma^2)$ populations, for σ varied from 0.1, 0.2, ..., 3.0. Figure 3.2 is a plot of each test's estimated power for various σ, based on 1000 replications of the simulation.

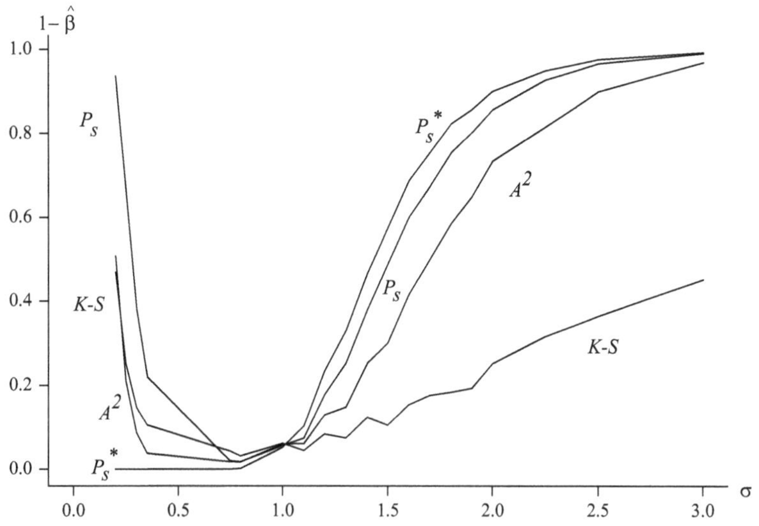

Fig. 3.2. Power functions for testing H_0: $X \sim N(0, 1)$ versus H_1: $X \sim N(0, \sigma^2)$ at level of significance $\alpha = 0.05$ using the K–S test statistic, A–D test statistic, and two test statistics based on the $\mathcal{P}$-vector

Figure 3.2 shows that:

- both P_s and $P_s^\star$ have significantly higher power than have A^2 and D_n for $\sigma > 1$,
- $P_s^\star$ slightly outperforms P_s for $\sigma > 1$,
- P_s clearly outperforms A^2 and D_n for $\sigma < 1$, and $P_s^\star$ has very low power.

The strong performance of P_s, four times the power of A^2 at $\sigma = 0.30$, for example, causes us to conjecture that this is a strong omnibus test statistic for guarding against departures from H_0. An estimate of the standard deviation of the estimate of a power value $1 - \beta$ is less than $\sqrt{(0.001)(\hat{\beta})(1 - \hat{\beta})}$, or about 0.015 for individual values plotted in the mid-height range of Figure 3.2. Thus, the improvement in power of P_s over A^2 shown in Figure 3.2 is indeed statistically significant. Because P_s seems to provide protection in both directions of departure of σ from 1, we decided to investigate the distribution of its critical points more fully.

Due to the dependence of the p_i elements, finding an analytic expression for the distribution of these statistics seems untenable; however, using simulation, we have approximated critical points for the P_s values shown in Table 3.1. The critical points of all statistics are those of the fully specified null distribution, case zero in D'Agostino's [40]. The simulation for this table relied on 10,000 iterations of each sample size, so only two digits are significant. The distribution of P_s seems to have a heavy right-hand tail, and as n increases, the tail becomes heavier. For $n \leq 25$ the three critical values shown in Table 3.1 increase in a nearly linear fashion.

3.6 Further Research

Coding up the procedures in a compiled language for execution speed would allow larger sample sizes to be considered. The APPL code used to compute the critical values in Table 3.1 is about 50 lines. Embedded in that code is the ability to find the n-fold vector of polynomials representing the $F_{Z_{(i)}}$, each of degree up to n, that make up the beta distributed CDFs used to calculate

Table 3.1. Estimated critical values for P_s at various sample sizes and levels of s-significance

n	$\alpha = 0.10$	$\alpha = 0.05$	$\alpha = 0.01$
2	4.9	6.1	8.9
3	7.6	9.1	13.4
4	10.1	12.1	17.0
5	12.6	15.3	21.5
6	15.1	18.1	24.4
7	17.7	21.1	28.2
8	20.4	23.9	32.0
9	22.7	26.8	36.5
10	24.9	29.4	39.5
11	27.9	32.2	43.7
12	30.0	35.2	48.0
15	37.5	44.0	59.6
20	50.7	58.7	81.1
25	63.2	76.2	116.5
30	80.0	107.1	218.4
40	445.0	576.5	776.8
50	1025.4	1108.8	1231.6

the $\mathcal{P}$-vector elements. The leverage in our approach through the $Z_{(i)}$ is that Maple creates and stores the CDFs. The same simulation in FORTRAN or C might be possible with public-domain subroutines that calculate a specific statistic, i.e., in this case p_i. One difficulty in this transition that the order statistics may have extremely large integers for the coefficients of the terms, as well as for their exponents. Maple is able to conduct the mathematical operations without representing these constants as floating point approximations. Thus, our precision exceeds an implementation in a high-level algorithmic language. We believe that the performance of P_s suggests attractive procedures for outlier detection and inferences with censored samples may be defined in terms of the $\mathcal{P}$-vector.

[illegible]

4

The "Straightforward" Nature of Arrival Rate Estimation?

Donald R. Barr, Andrew G. Glen, and Harvey F. Graf

Abstract How one estimates the parameter in a Poisson process depends critically on the rule used to terminate the sampling period. For observation until the kth arrival, or observation until time t, well-known maximum likelihood estimators (MLEs) can be used, although they can be biased if the sampling period is such that the expected number of arrivals is small. If one uses a stopping rule such as "observe until the kth arrival or time t," the form of the MLE becomes more complex. In the latter case, it appears a simple ad hoc estimator outperforms its MLE competitor.

Keywords Maximum likelihood estimator • Poisson process • Sampling plan

---

Originally published in *The American Statistician*, Volume 52, Number 4 in 1998, this work was one of the first papers that used APPL as its primary research tool. APPL was useful in finding many aspects that solidified this work, to include simulations of random processes, likelihood function calculations, and finding moments with the `Mean` and `Variance` procedures.

D.R. Barr
USMA, West Point, NY, USA

A.G. Glen (✉)
Department of Mathematics and Computer Science, The Colorado College, Colorado Springs, CO, USA
e-mail: andrew.glen.1984@gmail.com

H.F. Graf
US Army, Madison, AL, USA

A.G. Glen, L.M. Leemis (eds.), *Computational Probability Applications*, International Series in Operations Research & Management Science 247,
DOI 10.1007/978-3-319-43317-2_4

4.1 Introduction

One of the topics in an undergraduate probabilistic modeling course we teach is queuing. To maintain an applications focus we send students into the community to observe queuing systems and ask them to model those systems. Most students collect arrival and other data on simple queuing situations and attempt to fit models involving a Poisson arrival process. Initially, estimating the arrival rate for this process appeared straightforward, but as the exercise progressed, a debate developed concerning whether one should use an estimator based on mean interarrival times or one based on the observed number of arrivals. More specifically, suppose we observe a Poisson process with positive rate λ over a period of time t in which there were k arrivals, say at times $x_1, x_2, \ldots, x_k$. Two possible estimates of the arrival rate λ are:

$$\lambda_1 = \#\text{ arrivals/observed time} = k/t$$
$$\lambda_2 = 1/\text{mean interarrival time} = k/x_k, \quad \text{for } k > 0.$$

Consider the following example: Arrivals at a cashier are observed for a 10-min period. Three customers arrive, at times 1.5 min, 5.1 min, and 8.0 min, respectively. Because this concerns the number of arrivals in a fixed observation period, one would most likely estimate the arrival rate to be $\hat{\lambda}_1 = 0.3$ arrivals/min. If one had estimated arrival rate as the reciprocal of the mean interarrival time, the value $\hat{\lambda}_2 = 0.375$ arrivals/min would have been obtained. As we shall see, the second estimate is inappropriate in this case because it ignores the right-censored $(k+1)$st interarrival time.

As we attempted to provide guidance to our students, we realized that arrival rate estimation poses an interesting problem that requires careful consideration of the sampling plan. We surveyed several texts that address basic queuing principles and applications, such as Ross [140] and Cook and Russell [34], and found none discussed this estimation problem. It is addressed in more advanced monographs, such as Basawa and Rao [10], but these are somewhat inaccessible to undergraduate and beginning graduate students.

In what follows, we develop maximum likelihood estimators (MLEs) of the arrival rate λ in a Poisson process for each of three sampling plans. The first two cases, illustrated in the previous example, lead to the familiar estimators applied there; we compare performance of these estimators in the two plans. To avoid wasting our students' time, we had suggested they observe a selected queuing system for 30 min or until 30 arrivals occurred. This defines a third sampling plan that leads to a slightly more challenging estimation problem in which finding the MLE involves a numerical solution of the normal equation. We return to the problem of estimating arrival rate with this sampling plan (which we call "sampling plan 3") after first considering the two simpler plans and the two estimators mentioned above. A fourth plausible sampling plan would be to observe the process until both of the conditions (at least 30 arrivals in a time span of at least 30 min) are met, possibly to assure a "good set of data" is obtained. An approach analogous to the one we outline for sampling plan 3 could be applied in the latter case, but we do not pursue it here.

We use upper-case characters to represent random variables and corresponding lower-case characters to denote their outcomes (so an observation on the estimator $\hat{\Lambda}_1$ for the parameter λ gives the estimate $\hat{\lambda}_1$, for example).

There are many related applications areas. Feller [53] discussed the problem of estimating the size of an animal population from recapture data, in which animals are caught sequentially, tagged, and released. One can estimate population size using various sampling plans such as counting the number of catches until the kth time a tagged animal is caught, or counting the number of tagged animals observed in a total of n catches. Observing a Poisson process for a fixed period of time, as in the example above, gives a censored sample of interarrival times. Estimation of population characteristics with censored data is a common problem in reliability applications and survival analysis (see Cox and Oakes [35], for example).

Development and evaluation of estimators for use with these plans involves several statistical ideas and facts and is generally quite tractable mathematically. It also illustrates the point that choice of an estimator generally depends critically upon how the data were collected (in our case, on the sampling plan used). We believe it makes an excellent assignment for a senior- or early graduate-level course that includes statistical inference or data analysis.

4.1.1 Sampling Plan 1: Time Sampling

Sampling plan 1 calls for observing the Poisson process until the kth arrival, for k pre-selected. In this case, $T \sim \text{gamma}(k, \lambda)$ (see Parzen [129]), so the likelihood function is proportional to $\lambda^k e^{-\lambda t}$ and the MLE of λ is $\hat{\Lambda}_3 = k/T$. Because $t = x_k$, one would compute the same value for $\hat{\lambda}_1$ and $\hat{\lambda}_2$, so we can say either estimate is a value of $\hat{\Lambda}_3$ in this case, even though $\hat{\Lambda}_1$, $\hat{\Lambda}_2$, (defined in the following) and $\hat{\Lambda}_3$ are different random variables. Integrating with respect to the gamma density it follows that

$$
\begin{aligned}
E(\hat{\Lambda}_3) = kE(1/T) &= k \int_0^\infty \frac{1}{t} \frac{\lambda^k}{\Gamma(k)} t^{k-1} e^{-\lambda t} dt \\
&= \frac{k\lambda}{k-1} \int_0^\infty \frac{\lambda^{k-1}}{\Gamma(k-1)} t^{(k-1)-1} e^{-\lambda t} dt,
\end{aligned}
$$

and because the latter integrand is a gamma$(k-1, \lambda)$ density, we have $E(\hat{\Lambda}_3) = \lambda(k/k-1)$ so $\hat{\Lambda}_3$ is biased, seriously for small k. An unbiased estimator in this case could be obtained by "adjusting" $\hat{\Lambda}_3$, in the spirit used to obtain the usual adjusted MLE for σ^2 based on a random sample from a $N(\mu, \sigma^2)$ population. The resulting unbiased estimator, $\hat{\Lambda}_{3a} = \hat{\Lambda}_3(k-1)/k = (k-1)/T$, also has a smaller variance than $\hat{\Lambda}_3$. Indeed, for $k > 2$,

$$\begin{aligned} V(\hat{\Lambda}_{3a}) &= (k-1)^2 E(1/T^2) - E^2(\hat{\Lambda}_{3a}) \\ &= (k-1)^2 \int_0^\infty \frac{1}{t^2} \frac{\lambda^k}{\Gamma(k)} t^{k-1} e^{-\lambda t} dt - \lambda^2 \\ &= \frac{(k-1)^2 \lambda^2}{(k-1)(k-2)} \int_0^\infty \frac{\lambda^{k-2}}{\Gamma(k-2)} t^{(k-2)-1} e^{-\lambda t} dt - \lambda^2 \\ &= \lambda^2 \left[\frac{k-1}{k-2} - 1 \right] = \frac{\lambda^2}{k-2}. \end{aligned}$$

4.1.2 Sampling Plan 2: Count Sampling

Under sampling plan 2, the Poisson process is observed for a preselected period of time, t. Because $K \sim \text{Poisson}(\lambda t)$, it is easy to see the likelihood function is again proportional to $\lambda^k e^{-\lambda t}$ and it follows that $\hat{\Lambda}_1 = K/t$ is the MLE of λ in this case. One could alternately develop the MLE in terms of the censored sample of interarrival times as follows. We have observed an outcome on the interarrival times $X_1 - 0, X_2 - X_1, \ldots, X_k - X_{k-1}$, together with the censored interarrival time, $t - X_k$. One may ignore censoring of the first interarrival time (censored because we start observation some time after the arrival preceding the first in our sample), due to the memoryless property of the exponential distribution of interarrival times. (Feller [54] discussed this in the context of a "waiting time paradox.") The likelihood function is therefore the product of k exponential density values, times the exponential survivor function evaluated at the censored value:

$$L(\lambda) = \prod_{i=1}^{k} \lambda e^{-\lambda(x_i - x_{i-1})} \cdot \left(1 - \left[1 - e^{-\lambda(t - x_k)}\right]\right) = \lambda^k e^{-\lambda t},$$

and because this is proportional to the likelihood value obtained with the Poisson model, the same MLE results.

With data from count sampling, λ_1 is different from λ_2, and it is of interest to compare the performance of $\hat{\Lambda}_1$ with that of $\hat{\Lambda}_2 = K/X_K$ in this case. Without conditioning on $K > 1$, $\hat{\Lambda}_1$ is unbiased, because by the Poisson model, $\mathrm{E}(\hat{\Lambda}_1) = (1/t)E(K) = (1/t)\lambda t = \lambda$. Using similar reasoning, we see $\hat{\Lambda}_1$ has variance $V(\hat{\Lambda}_1) = \lambda/t$ in that case. However, given $K > 1$, $\hat{\Lambda}_1$ has expected value

$$E(\hat{\Lambda}_1) = \lambda \cdot \frac{1 - e^{-\lambda t}}{1 - e^{-\lambda t}\left[1 + \lambda t\right]},$$

which approaches λ from above as t increases, and variance

$$V(\hat{\Lambda}_1) = \frac{\lambda\left(1 - e^{-\lambda t}\left[1 + \lambda t\right]\right)\left(1 + \lambda t - e^{-\lambda t}\right) - \lambda t\left(1 - 2e^{-\lambda t} + e^{-2\lambda t}\right)}{t\left(1 - e^{-\lambda t}\left[1 + \lambda t\right]\right)^2}$$

which increases toward λ/t as t increases. Plots of these conditional moments of $\hat{\Lambda}_1$ are shown in Figure 4.1.

Properties of $\hat{\Lambda}_2$ are not so easily determined in this case, because both K and the sum of interarrival times X_K are random. We find the conditional mean of $\hat{\Lambda}_2$ given $K = k$ for $k > 1$, then take the expectation of that with respect to the conditional Poisson distribution of K, given $K > 1$.

It is well-known that, given $K = k$ arrivals in an interval $(0, t]$, the arrival times (the X's) are jointly distributed identically as k $U(0, t)$ order statistics (see Parzen [129], for example). Thus, X_k is distributed as t times a beta(k, 1) random variable Y, because the largest in a random sample of size k from a $U(0, 1)$ distribution is distributed beta(k, 1) (see Barr and Zehna [9], for example). With $Y \sim$ beta(k, 1), we have $f_Y(y) = ky^{k-1}$; $0 < y < 1$ so, assuming $k > 1$,

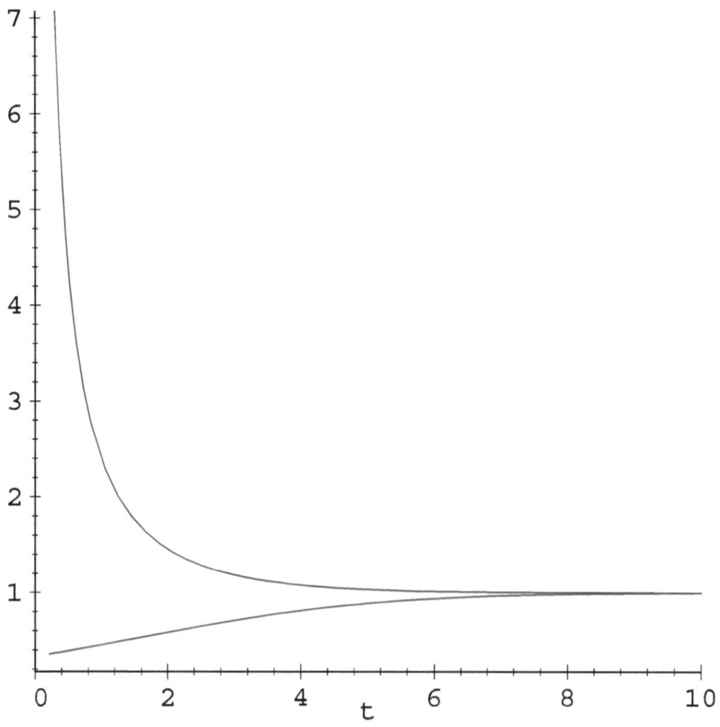

Fig. 4.1. Plot of the mean (*top curve*) and variance of Λ_1 (*bottom curve*), given $K > 1$, in the case $\lambda = 1$

$$E\left(\frac{1}{Y}\right) = k \int_0^1 \frac{1}{y} y^{k-1} dy = \frac{k}{k-1}$$
$$\Longrightarrow E(\Lambda_2 | K = k) = \frac{k}{t} \cdot \frac{k}{k-1}.$$

(The first result provides yet another opportunity to warn students about the difference between $E(1/Y)$ and $1/E(Y) = (k+1)/k$.)

We now "un-condition" by taking the expectation of $K^2/t(K-1)$ with respect to the Poisson distribution of K, conditioned by the event $K > 1$: for any fixed integer $n > 2$,

$$E_K\left[E(\hat{\Lambda}_2|K)|K>1\right] = \frac{1}{tP[K>1]}\sum_{k=2}^{\infty}\frac{k^2 m(k)}{k-1}$$
$$\cong \frac{1}{tP[K>1]}\left[\sum_{k=2}^{n-1}\frac{k^2 m(k)}{k-1} + \sum_{k=n}^{\infty} k m(k)\right],$$

in which $m(\cdot)$ denotes the Poisson (λt) mass function and the approximation involves replacing $k/(k-1)$ by 1 for k larger than $n-1$. Clearly, the approximation becomes exact as $n \to \infty$ and it will be good for sufficiently large n. The series in the approximation above can be expressed as $\lambda t - m(l) - \sum_{k=2}^{n-1} k m(k)$ so the approximate expression can be written in the form $\lambda A(\lambda t) + B(\lambda t, n)/t$, for $A(\lambda t) = P[K>0]/P[K>1]$, and

$$B(\lambda t, n) = \sum_{k=2}^{n-1}\frac{k m(k)}{(k-1)P[K>1]}.$$

Because $A(\lambda t) > 1$ and for any $n > 2$, $B(\lambda t, n) > P[1 < K < n]/P[1 < K] > 0$, it follows that $\hat{\Lambda}_2$ is biased. (This is also suggested by the fact that for any $k > 1$, $E(\hat{\Lambda}_2|K=k) = \hat{\lambda}_1 k/(k-1) > \hat{\lambda}_1$ and $\hat{\Lambda}_1$ is unbiased.) The error of the approximation is bounded above by the succession of increasing bounds

$$\left(\frac{n}{n-1}-1\right)\sum_{k=n}^{\infty} k m(k) < \frac{1}{n-1}(\lambda t - P[K<n]) < \frac{1}{n-1}\lambda t,$$

so the error is certainly less than σ when $n-1 > \lambda t/\sigma$. We computed numerical approximations for the case $\lambda = 1$, for $t = 0.5, 1.0, \ldots, 40.0$, using n values guaranteeing error less than 10^{-6}. A plot of these values is shown in Figure 4.2.

We observed that $A(\lambda t)$ converges rapidly down to 1.0 and for sufficiently large n, $B(\lambda t, n)$ converges fairly rapidly down to 1.0 as t increases, so for t greater than 6 or so, the bias is approximately $1/t$. [This is numerical confirmation of the fact that $\lim_{n\to\infty} B(\lambda t, n)$ approaches 1 from above as t increases.] We have plotted $1 + 1/t$ in Figure 4.2 for comparison purposes. This suggests one can get reasonable estimates of the bias of $\hat{\Lambda}_2$, for a given value of t and for parameter λ other than 1.0, by considering the values on the abscissa of Figure 4.2 to be values of λt (so if $\lambda = 2$ the bias for a plan with $t = 3$ can be read from Figure 4.2 at the value 6 on the abscissa, for example). A plot of these values is shown in Figure 4.2.

Again, one might use the adjusted estimator

$$\hat{\Lambda}_{2a} t/(t+1) \cdot \hat{\Lambda}_2 = tK/(t+1)X_K,$$

which would also reduce the variance. An argument along the lines used to approximate the mean of $\hat{\Lambda}_2$ might be used to get an approximation for its variance as well, but it seems to be mathematically intractable, so we turned to a Monte Carlo simulation to examine the mean and variance of the adjusted MLE, and to compare them with corresponding moments of $\hat{\Lambda}_1$. A summary of the results are shown in Table 4.1, in which estimates are conditioned on the event $[K > 1]$. In general, it appears the MLE is superior with this sampling plan, especially for small to moderate values of t.

Table 4.1. Mean and standard error of $\hat{\Lambda}_1$ and $\hat{\Lambda}_{2a}$ for count sampling, given $K > 1$; for $\lambda = 1$. MSE ratio is the ratio of the sample mean square error of $\hat{\Lambda}_1$ to that of $\hat{\Lambda}_{2a}$

t	Avg $\hat{\Lambda}_1$	Std dev $\hat{\Lambda}_1$	Avg $\hat{\Lambda}_{2a}$	Std dev $\hat{\Lambda}_{2a}$	MSE ratio	$P[K > 1]$
0.5	4.35	0.88	2.72	3.36	0.843	0.090
1.0	2.40	0.68	2.07	1.69	0.607	0.264
2.0	1.46	0.54	1.53	1.26	0.272	0.594
5.0	1.03	0.42	1.10	0.64	0.425	0.960
10.0	1.00	0.31	1.01	0.30	1.050	0.999
25.0	1.00	0.20	1.00	0.20	1.030	1.000

4.1.3 Sampling Plan 3: Limit Both Time and Arrivals

The sampling plan we unwittingly gave our students complicates arrival rate estimation. Our instructions (restated) were: for pre-selected limits $k^\star$ and $t^\star$, observe the system until the first occurrence of either of the events $[K = k^\star]$ or $[T = t^\star]$. With this stopping rule the process will pass through the point $(k^\star, t^\star)$ with probability zero, so there are two mutually exclusive and exhaustive cases to consider: (1) $1 < K < k^\star$ and $T = t^\star$; and (2) $K = k^\star$ and $T < t^\star$. Let us consider maximum likelihood estimation of λ for this sampling plan. The (conditional) likelihood function for each case is as follows:

Case 1. $1 < K < k^\star$ and $T = t^\star$.

This will occur with probability

$$\begin{aligned} p_1 &= P[\text{fewer than } k^\star \text{ arrivals as of time } t^\star | \text{at least two arrivals}] \\ &= \sum_{j=2}^{k^\star - 1} m(j; \lambda t^\star) / [1 - M(1; \lambda t^\star)], \end{aligned}$$

in which $m(j; \lambda t^\star)$ is the Poisson$(t^\star)$ mass function evaluated at j and M is the corresponding CDF. The conditional likelihood $m(k|l < K < k^\star)$ is, for integer k in the interval $(1, k^\star)$, the Poisson mass function divided by $p_1[1 - M(1; \lambda k^\star)]$.

Case 2. $K = k^\star$ and $T < t^\star$. This situation will occur with probability $p_2 = 1 - p_1$. The conditional likelihood, $f(t|T < t^\star)$, for $0 < t < t^\star$, is just the gamma $(k^\star, \lambda)$ density divided by $p_2[1 - M(1; \lambda t^\star)]$ because $[T < t^\star]$ is equivalent to $[k^\star$ or more arrivals by time $t^\star]$.

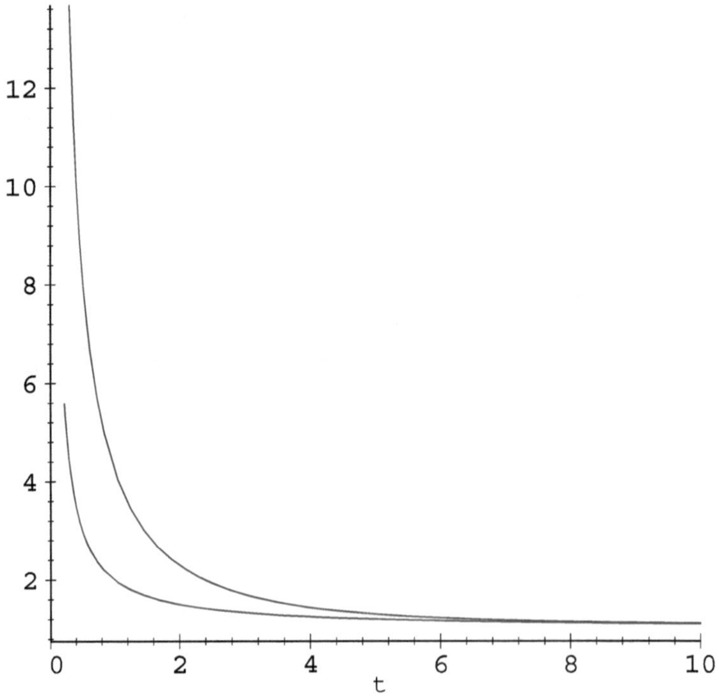

Fig. 4.2. Approximate Mean of Λ_2 (*top curve*) and $1+1/t$ (*bottom curve*), for $\lambda = 1$

The likelihood function $L(\lambda; k, t)$ can be found by multiplying each of the preceding conditional expressions by the corresponding probability p that it occurs, given at least two arrivals:

$$L(\lambda; k, t | k > 1) = \begin{cases} \frac{\lambda^k (t^\star)^k e^{-\lambda t^\star}}{\Gamma(k+1)[1-M(1;\lambda t^\star]} & \text{if } 1 < k < k^\star \\ \frac{\lambda^{k^\star} t^{k^\star - 1} e^{-\lambda t}}{\Gamma(k^\star)[1-M(1;\lambda t^\star]} & \text{if } 0 < t < t^\star. \end{cases}$$

We wish to maximize the first expression if Case 1 occurs, otherwise maximize the second expression. Equating the derivatives of these expressions with respect to λ to zero, we obtain the normal equation

$$e^{-\hat{\lambda} t^\star} = \frac{D}{\hat{\lambda}^2 t^{\star 2} + (1 - \hat{\lambda} t^\star) D}, \tag{4.1}$$

in which $D = k - \lambda t^\star$ if $(k, t^\star)$ is observed, and $D = k^\star - \lambda t$ if $(k^\star, t)$ is observed.

When the experiment is performed, we will observe either the point $(k^\star, t)$ or the point $(k, t^\star)$ (although, if $k < 2$, we perform another experiment), so to compute the value of the MLE, $\hat{\Lambda}_4$, it suffices to solve the normal equation (4.1), using the appropriate value of D. Note that the first expression with k set equal to its upper bound is equivalent to the second expression with t set equal to $t^\star$, so the two cases give the same estimate on the boundary $(k^\star, t^\star)$. Note also in Case 1, substitution of $\hat{\lambda}_1 = k/t^\star$ for $\hat{\lambda}$ in the normal equation gives the expression $e^{-k} = 0$. Thus, for k larger than 12 or so, $\hat{\lambda}_1$ is an excellent approximate solution. A similar comment holds for Case 2 and the approximate solution $\hat{\lambda}_3 = k^\star/t$.

We used numerical methods to find approximate solutions to these equations using the APPL running in Maple Version 4 software. We chose Maple because of its robust solving engine, using the `solve` command, which does not require initial estimates, unlike the secant method, for example. We created samples of $\hat{\lambda}_4$ of size 10,000 for several combinations of $t^\star$ and $k^\star$ (smaller sample sizes when $k^\star$ was too large for Maple to compute without crashing on our machine). We also calculated corresponding estimates $\hat{\lambda}_1 = k/t^\star$ and $\hat{\lambda}_{3a} = (k^\star - l)/t$ in each iteration. In addition, we calculated estimates, $\hat{\lambda}_5$, with an ad hoc estimator defined as follows:

$$\hat{\Lambda}_5 = \begin{cases} \hat{\Lambda}_1 & \text{if } (K, t^\star) \text{ is observed} \\ \hat{\Lambda}_{3a} & \text{if } (k^\star, T) \text{ is observed.} \end{cases}$$

This estimator is suggested by the relatively good performance of $\hat{\Lambda}_1$ with count sampling and $\hat{\Lambda}_{3a}$ with time sampling, and the idea that when $(K, t^\star)$ is observed the outcome is (conditionally) on a count sample, and similarly for $(k^\star, T)$ and time sampling.

In Table 4.2 we show the averages of 10,000 realizations of $\hat{\Lambda}_1$, $\hat{\Lambda}_{3a}$, $\hat{\Lambda}_4$, and $\hat{\Lambda}_5$, together with the corresponding estimated standard errors. To the right of the standard error data in Table 4.2 is a column showing the ratio of sample mean square errors of $\hat{\Lambda}_5$ and $\hat{\Lambda}_4$. We expected $\hat{\Lambda}_{3a}$ to perform well when there are few occurrences of Case 2, but it actually does fairly well in all cases shown in Table 4.2, except when there is a heavy preponderance of Case 1 outcomes. On the other hand, $\hat{\Lambda}_1$ performs poorly except when there is a heavy preponderance of Case 1 outcomes. This suggests $\hat{\Lambda}_5$, which "takes the best of both $\hat{\Lambda}_1$ and $\hat{\Lambda}_{3a}$," should perform well. Indeed, $\hat{\Lambda}_5$ outperforms the maximum likelihood estimator, $\hat{\Lambda}_4$, in all the cases we examined, as demonstrated in the "MSE ratio" column of Table 4.2. We see good asymptotic properties of the MLE showing up toward the bottom of the table, for $k^\star$ large. Together with its ease of application, the ad hoc estimator $\hat{\Lambda}_5$ would seem to be the preferred estimator for our third sampling plan.

4.2 Conclusions

Our attempt to avoid wasting students' time by suggesting the third sampling plan was reasonable, when an estimator appropriate for this case is used. For this sampling plan, it appears the ad hoc estimator $\hat{\Lambda}_5$ is generally superior, especially in light of the ease of computing its outcomes. It seems slightly simpler, and thus, possibly preferable, to use count sampling and the estimator $\hat{\Lambda}_1 = K/t$ or time sampling and the estimator $\hat{\Lambda}_{3a} = (k-1)/T$. As a practical matter, the estimates obtained with the latter estimators will be similar (differing by $1/t$) when x_k is not too small. In classroom discussionsof estimation

Table 4.2. Simulation comparison of several estimators with the plan "sample until $[T = t^\star$ or $K = k^\star]$." Ten thousand simulation replications (fewer when $k^\star > 10$) were made with $\lambda = 1$ for each $(k^\star, t^\star)$ case. The estimator $\hat{\Lambda}_4$ is the MLE for this case; $\hat{\lambda}_1 = k/t^\star$; $\hat{\lambda}_{2a} = (k^\star - 1)/t$; and $\hat{\Lambda}_5$ is an ad hoc estimator that takes value of $\hat{\lambda}_1$ when $(k, t^\star)$ is observed and value $\hat{\lambda}_{3a}$ when $(k^\star, t)$. "MSE Ratio" is the ratio of the estimated mean square error of $\hat{\lambda}_5$ and that of $\hat{\lambda}_4$

Plan		Mean				Std error				MSE	Case 1	Case 2
$k^\star$	$t^\star$	$\hat{\lambda}_1$	$\hat{\lambda}_{3a}$	$\hat{\lambda}_4$	$\hat{\lambda}_5$	$\hat{\lambda}_1$	$\hat{\lambda}_{3a}$	$\hat{\lambda}_4$	$\hat{\lambda}_5$	ratio	Count	Count
3	3	1.73	1.12	1.60	1.21	2.14	1.45	1.69	1.40	0.68	2778	7222
3	5	1.56	1.03	1.45	1.05	1.46	0.98	1.49	0.97	0.42	852	9148
5	3	1.30	0.93	1.19	1.19	0.75	0.67	0.81	0.56	0.48	7711	2289
5	5	1.23	0.96	1.19	1.04	0.79	0.66	0.82	0.60	0.54	4204	5796
5	5	1.23	0.95	1.19	1.04	0.74	0.61	0.76	0.55	0.52	4111	5889
5	10	1.25	1.00	1.22	1.00	0.71	0.57	0.73	0.57	0.61	303	9697
5	25	1.24	0.99	1.21	0.99	0.69	0.55	0.71	0.55	0.61	0	10,000
10	5	1.10	0.90	1.06	1.08	0.52	0.51	0.55	0.48	0.76	9270	730
10	10	1.07	0.95	1.07	1.00	0.43	0.40	0.43	0.36	0.70	4591	5409
10	25	1.12	1.01	1.12	1.01	0.40	0.36	0.40	0.36	0.81	0	10,000
25	5	1.03	0.83	0.98	1.03	0.35	0.35	0.38	0.35	0.87	7000	0
25	10	0.99	0.89	0.99	0.99	0.28	0.28	0.29	0.28	0.97	8000	0
25	25	1.02	0.98	1.02	1.00	0.19	0.19	0.19	0.17	0.83	3296	3704
25	30	1.04	1.00	1.04	1.00	0.16	0.15	0.16	0.15	0.87	773	4227
30	25	1.00	0.96	1.00	1.00	0.15	0.15	0.15	0.14	0.89	4101	899
30	30	1.02	0.98	1.02	1.00	0.15	0.15	0.15	0.14	0.84	2387	2613

of the intensity of a Poisson process, it is important to remind students to design the sampling plan in advance, and to adhere to the plan once data collection has begun. A seemingly inconsequential difference in deciding when to stop observing a Poisson process can make a big difference in how $\hat{\lambda}$ should be estimated.

5

Survival Distributions Based on the Incomplete Gamma Function Ratio

Andrew G. Glen, Lawrence M. Leemis, and Daniel J. Luckett

Abstract A family of probability distributions is constructed using the incomplete gamma function ratio. The family includes a number of popular univariate survival distributions, including the gamma, chi square, exponential, and half-normal. Examples that demonstrate the generation of new distributions are provided.

Keywords Probability • Univariate lifetime distributions

5.1 Introduction

The gamma function

$$\Gamma(\alpha) = \int_0^\infty e^{-t} t^{\alpha-1}\, dt$$

for $\alpha > 0$ is a generalization of the factorial function that is prevalent in probability and statistics. When the lower limit of the integral is replaced by x, the resulting function is defined as the incomplete gamma function

This original method for finding new probability distribution families relies heavily on APPL as its foundation for computational exploration. The use of APPL was pervasive in everything from creating the new distributions to finding their moments and classifying their hazard functions. It was also valuable to compare new distributions with known families of distributions to rule out redundancies.

A.G. Glen (✉)
Department of Mathematics and Computer Science, The Colorado College, Colorado Springs, CO, USA
e-mail: andrew.glen.1984@gmail.com

L.M. Leemis • D.J. Luckett
The College of William and Mary, Williamsburg, VA, USA
e-mail: leemis@math.wm.edu

A.G. Glen, L.M. Leemis (eds.), *Computational Probability Applications*, International Series in Operations Research & Management Science 247,
DOI 10.1007/978-3-319-43317-2_5

$$\Gamma(\alpha, x) = \int_x^\infty e^{-t} t^{\alpha-1}\, dt$$

for $\alpha > 0$ and $x > 0$. The incomplete gamma ratio

$$\frac{\Gamma(\alpha, x)}{\Gamma(\alpha)}$$

for $\alpha > 0$ and $x > 0$ is bounded below at 0 and bounded above at 1, as the numerator is always smaller than the denominator. Let $g(x)$ be a monotonic and increasing function that assumes non negative values on the interval $(0, \infty)$. Furthermore, assume that $\lim_{x \to 0^+} g(x) = 0$ and $\lim_{x \to \infty} g(x) = \infty$. Likewise, let $r(x)$ be a monotonic and decreasing function that assumes non negative values on the interval $(0, \infty)$. Furthermore, let g and r be differentiable on the interval $(0, \infty)$. Also, assume that $\lim_{x \to 0^+} r(x) = \infty$ and $\lim_{x \to \infty} r(x) = 0$. A family of survival distributions for a random variable X is generated with CDFs

$$F(x) = 1 - \frac{\Gamma(\alpha, g(x))}{\Gamma(\alpha)} \qquad x > 0 \tag{5.1}$$

for any $\alpha > 0$, and

$$F(x) = \frac{\Gamma(\alpha, r(x))}{\Gamma(\alpha)} \qquad x > 0$$

for any $\alpha > 0$. The conditions on $g(x)$ and $r(x)$ ensure that $F(x)$ will be a monotonically increasing function with $F(0) = 0$ and $\lim_{x \to \infty} F(x) = 1$.

The PDF for this family when $g(x)$ is specified is found by differentiating the CDF:

$$f(x) = F'(x) = \frac{d}{dx}\left(1 - \int_{g(x)}^\infty e^{-t} t^{\alpha-1} dt\right) \Big/ \Gamma(\alpha).$$

By the chain rule of differentiation, this reduces to

$$f(x) = e^{-g(x)} g(x)^{\alpha-1} g'(x) / \Gamma(\alpha) \qquad x > 0$$

for $\alpha > 0$. When $r(x)$ is specified, the PDF is found as

$$f(x) = F'(x) = \frac{d}{dx}\left(\int_{r(x)}^\infty e^{-t} t^{\alpha-1} dt\right) \Big/ \Gamma(\alpha)$$

which reduces to

$$f(x) = -e^{-r(x)} r(x)^{\alpha-1} r'(x) / \Gamma(\alpha) \qquad x > 0$$

for $\alpha > 0$.

This method of creating new distribution functions from this ratio of gamma functions has a very nice graphical representation. The integrand of the gamma function $b(x) = e^{-x}x^{\alpha-1}$ can be considered the base function of the ratio. It is a positive function and has finite area underneath it on the interval $(0, \infty)$. Thus, the denominator in Eq. (5.1) is the entire area under $b(x)$. The numerator is the right tailed area under $b(x)$ on the interval $(g(x), \infty)$. Clearly the ratio in Eq. (5.1) is bounded between 0 and 1, thus, it represents

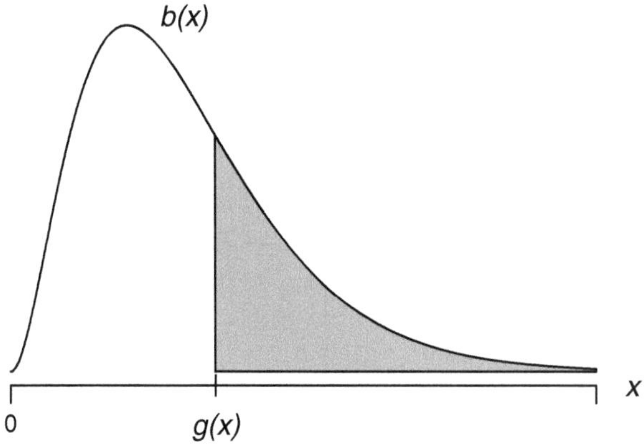

Fig. 5.1. The base function has finite area underneath it and the *shaded area* represents the portion of that area to the right of $g(x)$

the survivor function of a random variable. Figure 5.1 shows this relationship, in which the shaded area represents a portion of the entire area under the base, $b(x)$.

A number of popular lifetime distributions can be derived using PDFs of this form. Table 5.1 shows the appropriate $g(x)$ or $r(x)$ functions and α values associated with a number of PDFs, many of which are defined in Meeker and Escobar [113]. In the table and the analysis and examples that follow, let n be a positive integer and let α, λ, κ, and σ be positive parameters.

5.2 Properties and Results

The chosen function $g(x)$ or $r(x)$ is related to the hazard function of the resulting distribution. Let $f(x)$, $F(x)$, $S(x)$, and $h(x) = f(x)/S(x)$ be the population PDF, CDF, survivor function, and hazard function, respectively. These three results follow.

Theorem 5.1 Let the random variable X have PDF as defined above for a specified $g(x)$ and $\alpha = 1$. Then the hazard function $h(x) = g'(x)$, for $x > 0$, which implies the $g(x)$ is the cumulative hazard function for the distribution.

Proof. The hazard function of X is

$$h(x) = \frac{f(x)}{S(x)} = \frac{f(x)}{1 - F(x)} = \frac{e^{-g(x)}g'(x)}{\Gamma(1, g(x))} = \frac{e^{-g(x)}g'(x)}{\int_{g(x)}^{\infty} e^{-t}dt} = \frac{e^{-g(x)}g'(x)}{e^{-g(x)}} = g'(x)$$

for $x > 0$. □

Table 5.1. Parametric special cases

Distribution	$g(x)$ unless $r(x)$	α	$f(x)$
Exponential	λx	$\alpha = 1$	$\lambda e^{-\lambda x}$
Weibull	$(\lambda x)^{\kappa}$	$\alpha = 1$	$\lambda^{\kappa} x^{\kappa-1} \kappa\, e^{-(\lambda x)^{\kappa}}$
Rayleigh	$(\lambda x)^2$	$\alpha = 1$	$2\lambda^2 x e^{-(\lambda x)^2}$
Lomax	$\ln\left((\lambda x + 1)^{\kappa}\right)$	$\alpha = 1$	$\lambda\kappa/(1 + \lambda x)^{\kappa+1}$
Muth	$(1/\lambda)e^{\lambda x} - \lambda x - 1/\lambda$	$\alpha = 1$	$(e^{\lambda x} - \lambda)e^{-(1/\lambda)e^{\lambda x} + \lambda x + 1/\lambda}$
Half-normal	$x^2/(2\sigma^2)$	$\alpha = 1/2$	$\sqrt{2}\, e^{-x^2/(2\sigma^2)}/(\sigma\sqrt{\pi})$
Chi square	$x/2$	$\alpha = n/2$	$2^{-n/2} x^{n/2-1} e^{-x/2}/\Gamma(n/2)$
Erlang	x/λ	$\alpha = n$	$x^{n-1} e^{-x/\lambda}/(\lambda^n (n-1)!)$
Gamma	x/λ	–	$\lambda^{-\alpha} x^{\alpha-1} e^{-x/\lambda}/\Gamma(\alpha)$
Generalized gamma	$(x/\lambda)^{\kappa}$	–	$\kappa\lambda^{-\kappa\alpha} x^{\kappa\alpha-1} e^{-(x/\lambda)^{\kappa}}/\,\Gamma(\alpha)$
Inverse gamma	$r(x) = (\lambda x)^{-1}$	–	$\lambda^{-\alpha} x^{-\alpha-1} e^{-1/(\lambda x)}/\Gamma(\alpha)$

Theorem 5.2 Let the random variable X have PDF as defined above for a specified $r(x)$ and $\alpha = 1$. The hazard function $h(x) = -r'(x)/(e^{r(x)} - 1)$, for $x > 0$.

Proof. The hazard function of X is

$$h(x) = \frac{f(x)}{1 - F(x)} = \frac{-e^{-r(x)}r'(x)}{1 - \Gamma(1, r(x))} = \frac{-e^{-r(x)}r'(x)}{1 - \int_{r(x)}^{\infty} e^{-t}dt} = \frac{-e^{-r(x)}r'(x)}{1 - e^{-r(x)}} = \frac{-r'(x)}{e^{r(x)} - 1}$$

for $x > 0$. □

Comparing the well-known distributions in Table 5.1 to their hazard functions confirms these results. For example, the exponential distribution has a constant hazard rate $h(x) = \lambda$ for $x > 0$. The Weibull distribution has a hazard rate $\lambda^{\kappa}\kappa x^{\kappa-1}$ for $x > 0$. It is well-known that the Weibull distribution has an increasing failure rate when $\kappa > 1$ and a decreasing failure rate when $\kappa < 1$. This can be seen from the corresponding function $g(x) = (\lambda x)^{\kappa}$. When $\kappa > 1$, $g'(x)$ will be a monotonically increasing function.

These results tell us that a distribution specified by $g(x)$ and $\alpha = 1$ will have an increasing failure rate if $g''(x) > 0$ for all $x > 0$. The distribution will have a decreasing failure rate if $g''(x) < 0$ for all $x > 0$. Although the result in the $r(x)$ case is less tractable, similar conclusions can be made.

Next we consider taking transformations of random variables in this family. We will show that the proposed family of distributions is closed under certain 1–1 transformations. The following result holds in the case in which $g(x)$ is specified.

Theorem 5.3 Let the random variable X have PDF as defined above for a specified $g(x)$ and unspecified α. If $Y = \phi(X)$ is a 1–1 transformation from $\{x \mid x > 0\}$ to $\{y \mid y > 0\}$, then the PDF of Y is in the same family.

Proof. Let the random variable X have PDF

$$f_X(x) = e^{-g(x)} g(x)^{\alpha-1} g'(x) / \,\Gamma(\alpha) \qquad x > 0,$$

for $\alpha > 0$ and $g(x)$ for which the properties in Section 5.1 hold. Consider first the case of $Y = \phi(X)$ monotonically increasing. So $Y = \phi(X)$ is a 1–1 transformation from $\{x \mid x > 0\}$ to $\{y \mid y > 0\}$ with inverse

$$X = \phi^{-1}(Y)$$

and Jacobian

$$J = \frac{dX}{dY} = \frac{d}{dY}\phi^{-1}(Y).$$

By the transformation technique, the PDF of Y is

$$\begin{aligned} f_Y(y) &= f_X(\phi^{-1}(y)) \cdot J \\ &= \frac{e^{-g\left(\phi^{-1}(y)\right)} g\left(\phi^{-1}(y)\right)^{\alpha-1} g'\left(\phi^{-1}(y)\right)}{\Gamma(\alpha)} \cdot \frac{d}{dy}\phi^{-1}(y) \\ &= \frac{e^{-g\left(\phi^{-1}(y)\right)} g\left(\phi^{-1}(y)\right)^{\alpha-1}}{\Gamma(\alpha)} \cdot \frac{d}{dy} g\left(\phi^{-1}(y)\right) \qquad y > 0, \end{aligned}$$

because

$$\frac{d}{dy} g\left(\phi^{-1}(y)\right) = g'\left(\phi^{-1}(y)\right) \cdot \frac{d}{dy}\phi^{-1}(y)$$

by the chain rule of differentiation. This PDF can be recognized as being from the proposed family and is constructed using the $g(x)$ function $g\left(\phi^{-1}(y)\right)$. The case of $Y = \phi(X)$ monotonically decreasing is handled in a similar fashion. □

A similar result can be shown for the case in which $r(x)$ is specified. This means that the proposed family is closed under transformations such as, for example, $Y = \phi(X) = c_0 X^{c_1}$ for positive, real constants c_0 and c_1.

5.3 Examples

New distributions can be developed by defining $g(x)$ or $r(x)$ functions satisfying the conditions in Section 5.1 that may fit a particular purpose. Here we present three examples of new distributions. Each example displays different design features that make the new distributions desirable. While each could be further developed (as in publications such as Forbes et al. [55]), the following examples demonstrate how new distributions with certain useful properties are generated.

Example 5.1. A new life distribution with determinable hazard function properties can be derived. Let $g(x) = (\lambda x)^\kappa$, for $\lambda, \kappa > 0$. Let $\alpha > 0$ be unspecified. This creates a distribution with PDF

$$f(x) = \frac{\kappa e^{-(\lambda x)^\kappa} x^{\kappa\alpha - 1} \lambda^{\kappa\alpha}}{\Gamma(\alpha)} \qquad x > 0.$$

When $\alpha = 1$, this reduces to the Weibull(λ, κ) distribution. When $\alpha = 2$, this creates a "semi-Weibull" distribution with PDF of X

$$f(x) = \kappa \lambda^{2\kappa} x^{2\kappa - 1} e^{-(\lambda x)^\kappa} \qquad x > 0.$$

This PDF is quite similar to that of a Weibull(λ, κ) random variable. The hazard function in this case is

$$h(x) = \frac{\kappa \lambda^{2\kappa} x^{2\kappa - 1}}{(\lambda x)^\kappa + 1} \qquad x > 0.$$

The CDF and survivor function can also be obtained in closed form. This distribution will have a decreasing failure rate when $\kappa \le 1$ and an increasing failure rate when $\kappa \ge 1$. Another special case of this family occurs when we further assume that $\kappa = 1$. The PDF reduces to

$$f(x) = \lambda^2 x e^{-\lambda x} \qquad x > 0,$$

which is the PDF of an Erlang random variable with rate λ and $n = 2$ stages.

Example 5.2. New distributions can be created by expanding existing distributions with the α parameter. Consider the Lomax distribution with the additional parameter α. Let $g(x) = \ln(x + 1)$. This example introduces some likelihood calculations. The PDF is

$$f(x) = \frac{1}{\Gamma(\alpha)(x+1)^2} \left[\ln(x+1)\right]^{\alpha - 1} \qquad x > 0.$$

The likelihood function is

$$L(\alpha)=\prod_{i=1}^{n}f(x_i)=[\Gamma(\alpha)]^{-n}\left(\prod_{i=1}^{n}\frac{1}{(1+x_i)^2}\right)\left[\prod_{i=1}^{n}\ln(1+x_i)\right]^{\alpha-1}$$

and the log likelihood function is

$$\ln L(\alpha)=-n\ln\bigl(\Gamma(\alpha)\bigr)-2\sum_{i=1}^{n}\ln(1+x_i)+(\alpha-1)\sum_{i=1}^{n}\ln\bigl(\ln(1+x_i)\bigr).$$

The score is

$$\frac{\partial \ln L(\alpha)}{\partial\alpha}=-\frac{n\Gamma'(\alpha)}{\Gamma(\alpha)}+\sum_{i=1}^{n}\ln\bigl(\ln(1+x_i)\bigr).$$

The equation

$$\frac{\Gamma'(\alpha)}{\Gamma(\alpha)}=\frac{1}{n}\sum_{i=1}^{n}\ln\bigl(\ln(1+x_i)\bigr)$$

must be solved numerically to compute the maximum likelihood estimator $\hat{\alpha}$ for a particular data set.

Example 5.3. New distributions with closed-form moments are possible. Let $g(x)=\lambda x^{1/\alpha}$. This yields the PDF

$$f(x)=\frac{\exp(-\lambda\,x^{1/\alpha})\lambda^{\alpha}}{\Gamma\,(\alpha+1)}\qquad\qquad x>0.$$

This distribution has mean

$$\mu=\frac{\lambda^{-\alpha}4^{\alpha}\Gamma\,(\alpha+1/2)}{2\sqrt{\pi}}$$

and variance

$$\sigma^2=\frac{\lambda^{-2\alpha}\left(4\pi\Gamma\,(3\alpha)-16^{\alpha}\bigl(\Gamma\,(\alpha+1/2)\bigr)^2\Gamma\,(\alpha)\right)}{4\pi\Gamma\,(\alpha)}.$$

Furthermore, all integer moments about the origin can be calculated with the moment function

$$E(X^n)=\int_0^{\infty}x^n f(x)\,dx=\frac{\lambda^{-n\alpha}\Gamma\,(\alpha+\alpha n)}{\Gamma\,(\alpha)}.$$

5.4 Conclusions

An unlimited number of survivor distributions can be generated using the incomplete gamma function ratio. Several popular survivor distributions, such as the Weibull and gamma distributions, are included in this class. The class is closed under monotonic transformations. Future research is showing that any base function that is positive and has finite area can be used to generate further new distributions.

6

An Inference Methodology for Life Tests with Full Samples or Type II Right Censoring

Andrew G. Glen and Bobbie L. Foote

Abstract We present a methodology for performing statistical inference procedures during the actual conduct of a life test experiment that can reduce time on test and cost. The method relies on properties of conditional order statistic distributions to convert censored data into iid uniform random variables. A secondary result presents a new test for uniformity based on the convolution of these iid uniform random variables that is higher in power than the benchmark Anderson–Darling test statistic in certain cases.

Keywords Computational algebra systems • Goodness-of-fit • Probability integral transform • Uniformity testing

---

Originally published in *IEEE Transactions on Reliability*, Volume 58, Number 4, in 2009, this paper is one of best uses of APPL in the conduct of the research as well as in the final result. A new procedure in APPL calculates the goodness-of-fit test statistic, its distribution under H_0, and the exact p-value of the test statistic. The APPL procedures `OrderStat` and `Truncate` enable the derivation of conditional order statistics that transform into iid uniform random variables under the null hypothesis. This algorithm was granted a patent for its innovative use.

A.G. Glen
Department of Mathematics and Computer Science, The Colorado College, Colorado Springs, CO, USA
e-mail: andrew.glen.1984@gmail.com

B.L. Foote (✉)
USMA, West Point, NY, USA

A.G. Glen, L.M. Leemis (eds.), *Computational Probability Applications*, International Series in Operations Research & Management Science 247,
DOI 10.1007/978-3-319-43317-2_6

Notation

n	Sample size
r	Right censored item number of highest observation
$X_{(25:10)}$	The tenth order statistic from a sample of 25 items
$F_{X_{(n:i)}}$	The CDF of the ith order statistic from a sample of size n
T_r	The T statistic for a r known order statistics of a lifetest
μ	The expected value of a random variable
σ^2	The variance of a random variable

Acronyms

PIT	Probability Integral Transformation
A–D	Anderson–Darling
APPL	A Probability Programming Language

6.1 Introduction and Literature Review

Life test experiments are designed to gain an understanding of the probabilistic properties of the lifetime of, for example, a new electrical circuit, a drug treatment, a mechanical component, or a system of components. Often, the costs of life tests, in both time and money, constrain the design of the experiment, limiting the number of items placed on test and the length of the test. Many times the length of the experiment cannot be estimated accurately in advance and often one is faced with censored data in an ongoing experiment. For such cases, we propose a methodology that gives exact statistical inference on full or censored samples. Further, while this methodology can be combined with existing test statistics, such as the Anderson Darling (A–D), we introduce a new statistic that has higher power than the A–D when the true mean decreases.

Consider a system, process, or component with a fully specified lifetime CDF $F(x)$. Should an improved system, process, or component come along, both producers and consumers would like to verify that the new item is better than the existing item, often by determining if its mean lifetime has improved. In the life test, it would be highly desirable to stop the test when enough evidence exists to support either claim. Such censoring, commonly called Type I (stop after time t) or Type II (stop after r items fail), can produce statistical inference.

We propose a methodology that will allow for Type II right censoring in the design and conduct the life test or a complete sample if the data are available. If, for example, one could afford a life test with $n = 10$ items to fail, a certain level of statistical power could be achieved if the test continued until the completion of n failures. Consider, however, an example in which

$n = 25$ items are placed on test with $r = 10$ as the designated censoring value. Obviously, the second test would conclude more quickly, as the expected time on test would be the mean failure time of $X_{(25:10)}$, the tenth order statistic from a sample of 25 items, under the null hypothesis. Now, consider a slightly different example, where $n = 25$ items are placed on test. Experimenters note that, after $r = 3$ failures, lifetimes seem to be substantially longer than the original system. After $r = 6$ failures, they are convinced, at least anecdotally, that the new system has a longer expected lifetime. It is desirable to gain inference at each r value, as it may be possible to terminate the life test with an early, satisfactory statistical result.

What follows is a new methodology that will allow for instantaneous assessment of the life test at every failure. We rely on properties of conditional order statistic distributions to provide inference for censored data. We further rely on the advances of computer algebra systems, especially APPL (Glen et al. [60], as well as Drew et al. [46]), as our technique requires calculating many CDFs of conditional order statistics and possibly the distribution of the test statistics. The method we propose transforms either a complete data set or a right-censored data set, via two probability integral transformations (PIT) and conditional order statistics, into an un-ordered, iid sample of uniformly distributed data on the open interval (0,1), which we abbreviate $U(0,1)$. At this point, one may then apply any test of uniformity to gain inference. We next investigate the test statistic T_r, based on the sum of r iid $U(0,1)$ random variables, which can result in higher power than the A–D statistic when finding differences in the mean of the item in question. As an outcome, experimenters may purposefully design life tests with higher values of n so that the test can be terminated early, thus, saving time, money, and items that were destroyed during the testing.

Obtaining statistical inference while reducing time on test has been an ongoing research topic. Nelson [117, 118] investigated the idea of "precedence testing" which compares two samples in a non-parametric setting to determine which sample appears to be preferred and can give early indications of the outcome. More recently, Wu et al. [174] extended the research of Balasooriya [6] to obtain statistical inference from censored samples. Wu et al. expanded the research to the Weibull distribution from the two-parameter exponential distribution. Our results adopt a different approach and are applicable for most continuous distributions, not just the Weibull distribution. Testing for uniformity of a sample has many applications, many of which are explained in Chap. 8 of D'Agostino and Stephens [40]. Rosenblatt [142] presents theory that transforms joint conditional random variables to ordered, uniformly distributed random variables for the censored case (we will instead transform censored data to a complete un-ordered set of iid uniform data). David [43] discusses the Markov nature of conditional order statistics. He explains a technique that equates conditional order statistics with specific truncated order statistics, a result that we will use as part of our method. O'Reilly and Stephens [125] use a Rosenblatt transform, then invert that transformed data to test ordered uniform data (we will create un-ordered uniform data). Hegazy and Green [66] present work on goodness-of-fit using expected values of order statistics with approximations used for critical values. Evans et al. [52] present exact distributions for order statistics for discrete random variables. Leemis and Evans [94] also worked with convolutions of discrete random variables to produce exact distributions. In both these articles, the authors make use of computer algebra systems in a similar way that we do for exploring new statistical procedures. Michael and Schucany [114] also present a transformation that takes censored data and transforms it into ordered uniform data. They, as well as Stephens [152], point out that the A–D statistic is generally more powerful than the other well-known goodness-of-fit statistics in the case when the mean has shifted. Thus, we will compare the power of T_r with only that of A–D to show higher power than A–D in detecting decreases in the mean. Additionally, this method was approved for a patent by the US Patent Office, a government held patent attributed to the authors, number 7010463 (US Patent Office [162]). The authors approve of the free, noncommercial use of this methodology by the reader.

6.2 The Methodology for Censored Data

Assume that the lifetime of an existing item is distributed by the all-parameters-known continuous random variable X with CDF $F(x)$. Let n items be on life test and let r be the Type II right censoring order statistic number. Recall that, in a life test, failure data arrive in increasing magnitude; in other words, in the form of order statistics. The ordered lifetime data $x_{(n:i)}$ have CDFs from their appropriate order statistics $F_{X_{(n:i)}}(x_{(n:i)})$, $i = 1, 2, \ldots, r$, (note $X_{(n:i)}$ will now be abbreviated $X_{(i)}$). In his work on order statistics, David and Nagaraja [43, p. 20] explains two useful properties that we employ: that order statistics form a Markov chain, in that for $r < s$,

$$f_{X_{(s)}|X_{(r)}=x_{(r)},X_{(r-1)}=x_{(r-1)},\ldots,X_{(1)}=x_{(1)}}(y) = f_{X_{(s)}|X_{(r)}=x_{(r)}}(y);$$

and that deriving the distribution of these order statistics is made simpler with truncated distributions. Theorem 2.7 on the same page of David and Nagaraja's text explains "For a random sample of n from a continuous parent, the conditional distribution of $X_{(s)}$, given $X_{(r)} = x$ $(s > r)$, is just the distribution of the $(s-r)$th order statistic in a sample of $n-r$ drawn from $f(y)/[1-F(x)]$ $(y \geq x)$, i.e., from the parent distribution truncated on the left at x." Thus, for our purposes, the CDF of the ith order statistic, given the $(i-1)$th data point, $F(x_{(i)}|x_{(i-1)})$, is that of the random variable $X_{(1)}$ with support $x_{(i-1)} < x_{(i)} < 1$. David and Nagaraja show this is the first order statistic from a sample size $n-(i-1)$ from the parent distribution of X truncated on the left at $x_{(i-1)}$. In other words, the distribution is independent of $x_{(1)}, x_{(2)}, \ldots, x_{(i-3)}$, and $x_{(i-2)}$; and is therefore memoryless. Because each of the conditional distributions can be computed, conducting separate PITs on each data value, $F_{X_{(i)}|X_{(i-1)}}(x_{(i)})$, $i = 2, 3, \ldots, r$ will produce a sample of r iid $U(0,1)$ random variates, see Rosenblatt [142, p. 470] to which any uniformity test can be applied. We estimate power curves for T_r and the A–D test. In cases where the mean decreases from μ_0, the test based on T_r has higher power. In cases where the mean increases from μ_0, the A–D has higher power. As shown in the next section, T_r is defined as

$$T_r = \sum_{i=1}^{r} F_{X_{(i)}|X_{(i-1)}}(x_{(i)}),$$

where $F_{X_{(1)}|X_{(0)}}$ is defined to be $F_{X_{(1)}}$ and r is the size of the censored sample. The distribution of T_r is that of the sum of r iid $U(0,1)$ random variables.

6.3 The Uniformity Test Statistic

The test statistic we propose has the distribution of the convolution of iid $U(0,1)$ random variables and can be written in two forms:

$$T_n = \sum_{i=1}^{n} U_i = \sum_{i=1}^{n} F_X(X_i)$$

for a complete data set and

$$T_r = \sum_{i=1}^{r} F_{X_{(i)}|X_{(i-1)}}(x_{(i)})$$

for a Type II right censored sample. Prior to settling on this statistic, we explored other functions of iid $U(0,1)$ random variables. One option we explored was finding the distribution of $C = \sum_{i=1}^{n} \csc(U_i)$, as the cosecant function magnifies the statistic when the tails are too heavy. The magnification happens at a quicker rate than that of $-\ln(U)$, (the basis of A–D test statistic) and we found this statistic had slightly higher power than A–D, when testing for shifts in σ_a away from σ_0. The statistic had appreciably less power, though, when testing for changes in μ, a fact geometrically understandable, as the changes in μ do not exaggerate the test statistic. Furthermore, the exact distribution of C could not be found and critical points had to be estimated with Monte Carlo simulation, an inconvenience we wanted to avoid. We also considered $\min(U_1, U_2, \ldots, U_n)$ and $\sum_{i=1}^{n} \tan(U_i)$ and found similar results.

We found success with the test statistic $T_n = \sum_{i=1}^{n} U_i$ as a test for uniformity. Finding the distribution of the convolution of n iid $U(0,1)$ random variables is calculable; see, for instance, Johnson et al. [72]. We have also derived (in APPL) the exact distributions of T_n for reasonable sample sizes, currently $n \leq 50$. An example of a complete PDF for the distribution of T_7, for example, is

$$f(x) = \begin{cases} \frac{1}{720}x^6 & 0 < x < 1 \\ \frac{7}{120}x - \frac{7}{48}x^2 + \frac{7}{36}x^3 - \frac{7}{48}x^4 + \frac{7}{120}x^5 - \frac{1}{120}x^6 - \frac{7}{720} & 1 < x < 2 \\ \frac{1337}{720} - \frac{133}{24}x + \frac{329}{48}x^2 - \frac{161}{36}x^3 + \frac{77}{48}x^4 - \frac{7}{24}x^5 + \frac{1}{48}x^6 & 2 < x < 3 \\ -\frac{12089}{360} + \frac{196}{3}x - \frac{1253}{24}x^2 + \frac{196}{9}x^3 - \frac{119}{24}x^4 + \frac{7}{12}x^5 - \frac{1}{36}x^6 & 3 < x < 4 \\ \frac{59591}{360} - \frac{700}{3}x + \frac{3227}{24}x^2 - \frac{364}{9}x^3 + \frac{161}{24}x^4 - \frac{7}{12}x^5 + \frac{1}{48}x^6 & 4 < x < 5 \\ -\frac{208943}{720} + \frac{7525}{24}x - \frac{6671}{48}x^2 + \frac{1169}{36}x^3 - \frac{203}{48}x^4 + \frac{7}{24}x^5 - \frac{1}{120}x^6 & 5 < x < 6 \\ \frac{117649}{720} - \frac{16807}{120}x + \frac{2401}{48}x^2 - \frac{343}{36}x^3 + \frac{49}{48}x^4 - \frac{7}{120}x^5 + \frac{1}{720}x^6 & 6 < x < 7\,. \end{cases}$$

The distributions of T_n for $n \le 50$ are available from the first author. Because the distribution function of T_n is known, exact critical values and p-values are calculable. Tables of critical values used for our power simulations are available from the first author and are left out of this article for brevity. Furthermore, as APPL can find exact p-values for these distributions, tables such as these are becoming less necessary. As an aside, it is well known that convolutions of uniform random variables tend quickly toward normality as n increases. Thus, it becomes less necessary to find the exact distribution of the statistic well before $n = 50$ as normal approximations are virtually identical. However, for lower values of n (and r, as we will show in the next section), it is important to have a way to calculate the exact p-values using APPL, or some other algorithm.

6.4 Implementation Using APPL

The theory of the method and the statistic are straightforward; however, due to the need for multiple exact distribution functions that have support values which depend on the data, the implementation is made practicable only with automated probabilistic software. We implement the new method and new statistic in APPL for a number of reasons. APPL allows for the use of any continuous distribution (well-known distributions, as well as ad hoc) to specify the null hypothesis and conducts the necessary PITs for these distributions. APPL calculates the CDFs of order statistics, as well as truncated distributions. In effect, APPL-based Monte Carlo simulations allow for random functions to be simulated. APPL also has the distributions for sums of iid uniform random variables already computed, so exact p-values are calculable. The algorithm for the methodology follows.

- Specify the lifetime distribution of the existing (old) system, $F(x)$.
- During the life test experiment, note n and create the vector of r observed failure times.
- Calculate $z_{(i)} = F(x_{(i)})$, $i = 1, 2, \ldots, r$, which are distributed as the first r order statistics out of a sample of n $U(0,1)$ random variables.
- Calculate the un-ordered, iid $U(0,1)$ (under the null hypothesis) $u_i = F_{Z_{(i)}|Z_{(i-1)}}(z_{(i)})$, $i = 1, 2, \ldots, r$. Note: we perform the PIT with $F(x)$ and then conduct the conditional order statistics PIT using the uniform

conditional order statistic distributions instead of the other way around. These two ways have been shown to be equivalent [59], but this method is preferred because the conditional order statistics of the uniform distributions are much more tractable than conditional order statistics using the parent distribution F. Also, note that we find the conditional order statistic using the truncation of the parent distribution method outlined by David and Nagaraja [43].

- Sum the u_i values to get the T_r statistic, or use any test of uniformity on the u_i values.
- Calculate the p-value with twice the area of the outer tail of the T_r distribution, or use tables for the A–D statistic. For T_r, we set the p-value to two times the probability of the outer tail in order to guard against shifts in the mean in both directions. The statistic is obviously symmetric and is treated much like a two-tailed Student t test statistic when comparing to the omnibus A–D. One tailed tests can also be performed, but clearly not compared to the two-tailed A–D.

The APPL code that implements this algorithm to calculate the statistic is as follows.

```
> # take the r censored values in 'data' and PIT them into the
> # list 'Zdata'
> for i from 1 to r, do
>   Zdata := [op(Zdata), CDF(Nulldist, data[i])];
> od;
> # tt sum the independent uniforms to for the statistic 't_stat'
> # starting with the first failure
> t_stat:=CDF(OrderStat(U(0, 1), n, 1), Zdata[1]);
> # then add up the subsequent failures until r is reached
> if (r > 1), then
>   for i from 2 to r, do
>     t_stat := t_stat + CDF(OrderStat(Truncate(U(0, 1),
>       evalf(Zdata[i-1]), 1), n - (i - 1), 1), Zdata[i]);
>   end do;
> end if;
> Tr_distn := cat('T',r);
> # now return the statistic, the lower and upper tail p-values
> # using APPL's 'CDF' command
> RETURN(t\stat, 2 * CDF(Tr_distn, t_stat),
>   2 * (1 - CDF(Tr_distn, t_stat)));
```

This algorithm is implemented in a new APPL procedure called `CensoredT` and its use is illustrated in appendix of this paper. The code is available from the first author.

6.5 Power Simulation Results

In this section, we discuss the results of power simulations that compare the T_r statistic to the A–D test statistic using a level of significance of $\alpha = 0.05$ for data from the normal, Weibull and gamma distributions. We also rely on previous power studies by Stephens [152] and Michael and Schucany [114] that establish the A–D test statistic as generally more powerful than other statistics in testing for uniformity, especially when detecting a shift in μ. These other statistics include the Kolmogorov–Smirnov D, D^+, and D^- test statistics, the Cramér–von Mises W^2 test statistic, the Kupier test statistic, and the Watson U^2 test statistic. As the A–D test statistic is generally more powerful than these statistics, we opine that it is sufficient to benchmark the new T_r test statistic against the A–D test statistic. As our intended purpose for the test statistic is to determine if the life testing indicates changes in mean lifetime, this power simulation varies μ_1 from μ_0, but fixes the variance. In every case, for the normal, gamma, and Weibull distributions, we fixed $\sigma_0^2 = 10$ and only varied μ above and below $\mu_0 = 100.0$. Thus, the hypotheses considered are

$$H_0 : \mu = 100 \qquad \text{versus} \qquad H_1 : \mu \neq 100.$$

For clarity, the parameterization of the Weibull distributions uses PDF $f(x) = \kappa\lambda^\kappa x^{\kappa-1}e^{-(\lambda x)^\kappa}$ for $x, \kappa, \lambda > 0$. The gamma distributions uses the parameterization with PDF $f(x) = \alpha\,(\alpha x)^{\beta-1}\,e^{-\alpha x}/\Gamma\,(\beta)$ for $x,\ \alpha,\ \beta, > 0$. Parameter values used to generate the variates from each of these distributions were found by solving the 2×2 set of equations comprised of setting the expression of the mean equal to 100 and the expression for the variance equal to 10. The actual parameters for the Weibull and gamma distributions that correspond to the simulations in Figures 6.1, 6.2, 6.3 are presented in Table 6.1. Those of the normal distributions are self-evident from the moments. Figures 6.1, 6.2, 6.3 show the power curves for both the T_r test statistic (shown as a solid curve) and A–D test statistic (shown as a dashed curve). In each plot, the horizontal axis is r, the right censored number. The vertical axes are the simulated power curves. Note in Figures 6.1, 6.2, 6.3, the T_r outperforms the A–D whenever the data were simulated from a distribution with $\mu < \mu_0$. This simulated result holds true for all values of r, except for some early values of $r < 5$. However, in the cases where $\mu_0 < \mu$, the A–D test outperformed the T_r test for all three families of distributions. Thus, we see that T_r and A–D compete with one another depending on the direction of the shift in the mean. Each simulation created 2000 samples for each parameter value and both T_r and A–D were computed at each iteration. One may estimate the standard deviation of each power estimate with $\sqrt{\hat{\beta}(1-\hat{\beta})/2000}$, which is approximately 0.01 for mid-values in each figure. Most differences between the pairs of power estimates of T_r test statistic and A–D test statistic are multiple standard deviations apart. Furthermore, because the power of T_r test statistic is higher than the power of A–D test statistic in all cases

Table 6.1. The simulations shown in Figures 6.1, 6.2, 6.3 were created from distributions with the following parameters that match the desired μ and σ^2 values in the two left hand columns

Normal		Weibull		Gamma	
μ	σ^2	λ	κ	α	β
97.5	10	0.01011	38.832	9.75	950.625
98.0	10	0.01005	39.034	9.80	960.400
98.5	10	0.01000	39.237	9.85	970.225
99.0	10	0.00995	39.440	9.90	980.100
99.5	10	0.00991	39.643	9.95	990.025
null					
100.0	10	0.00986	39.845	10.00	1000

where $r > 5$ for all three families at all μ values, a simple signs test is statistically significant evidence of the higher power of the T_r test statistic. Thus, the improvement of T_r test over A–D test is statistically significant for all but the earliest values of r when $\mu < \mu_0$. Many different distributions were investigated for various families, parameter values, sample sizes (n values) and censoring values (r). What we presented is representative of what we have found in general. The particular families, parameters, and censoring values were chosen so that power curves can be shown graphically, which we opine as more instructive than a typical tabular presentation of power values.

An important note involves the complexity of the Monte Carlo simulations. The simulations require, among other things, the CDF of the parent distributions (non-trivial in the case of the gamma), the ability to conduct a PIT on the data, the exact CDFs of truncated order statistics created from the data, and, when using T_r, the exact distribution of the test statistic. None of this is practicable outside of a computer algebra system, certainly not in well-known statistics software packages. APPL enables researchers to simulate with functions of random variables, not just random numbers, instantaneously creating CDFs that only get used once. To see some of these distributions and to get an appreciation for the simple case of $r = 5$ refer to the example in the appendix.

6.6 Some Applications and Implications

This methodology has potential for significant advances in reliability engineering life testing, pharmaceutical drug tests, or any sort of experiment where data come naturally in ordered form. The sequential testing ability allows for a test to be terminated early, hence ending an expensive experiment. Other implications of this research are as follows.

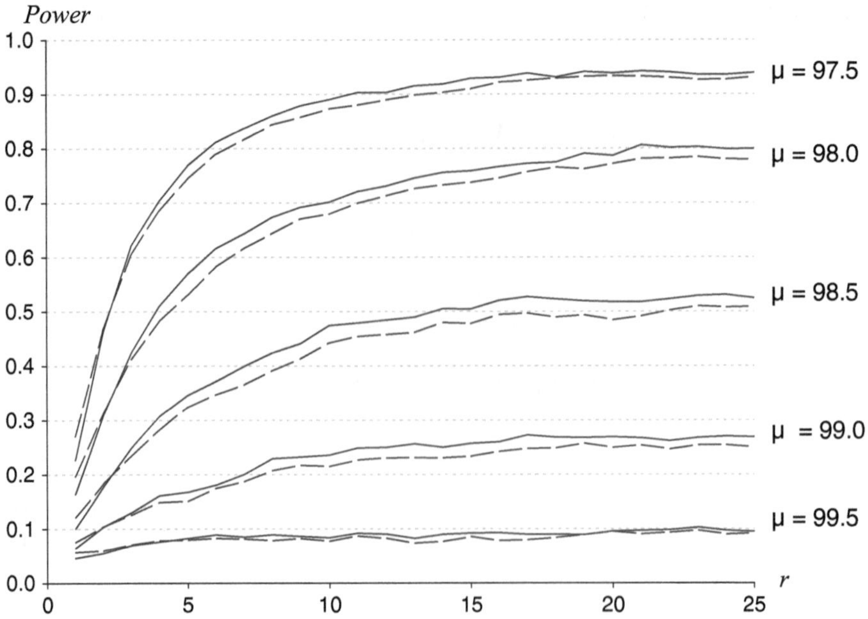

Fig. 6.1. Power curves displaying the results of Monte Carlo power simulation with underlying normally distributed data, $\sigma^2 = 10$, $\alpha = 0.05$ and $n = 25$. Under H_0, $\mu = 100$. The *solid line* is the T_r power curve and the *dashed line* is the A–D power curve. Note, as the true value of μ decreases from 100, the test based on the T_r test statistic outperforms the test based on the A–D test statistic

- Good statistical power for censored samples is possible for a wide range of experiments.
- Experiments can be designed with intentionally large values of n, knowing that they could stop at a (possibly predetermined) relatively small value of r.
- Experiments can be tracked in real-time to determine whether a pattern of p-values that indicates enough evidence of improvement has been attained.
- Computer algebra systems are adept enough for computing the multiple CDFs needed for such a statistic and methodology to be practical in its implementation.
- Creating Monte Carlo simulations with functions of random variables, not just random numbers, is possible in the APPL environment.

6.7 Conclusions and Further Research

A new life test methodology and a new uniformity test statistic have been developed, tested, and presented. Increases in power have been found compared to the benchmark A–D test statistic in cases where the true mean decreased in

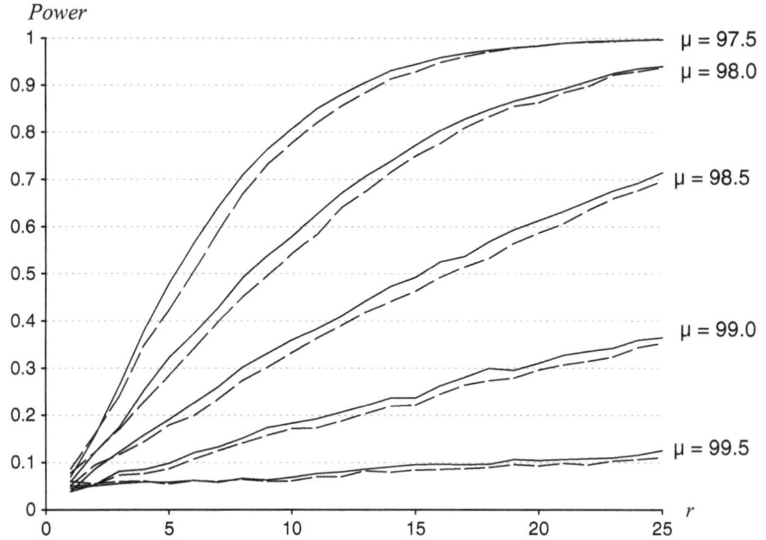

Fig. 6.2. Power curves displaying the results of Monte Carlo power simulation with underlying Weibull distribution of data, $\sigma^2 = 10$, $\alpha = 0.05$ and $n = 25$. Under H_0, $\mu = 100$. The *solid line* is the T_r power curve and the *dashed line* is the A–D power curve. Note, as the true value of μ decreases from 100, the test based on the T_r test statistic outperforms the test based on the A–D test statistic

value. Additionally, exact p-values for this test statistic are achievable. Also, relatively high power is achieved using the T_r test statistic on censored samples, allowing for life tests to be terminated early. Further research is needed to investigate how high to set n and r in experimental designs to gain possible advantages in lower time on test, lower cost, and fewer failed items as a result of the experiment. For example, if a budget can afford 25 items failing, perhaps it would be more effective to put 50 items on test, knowing ahead of time that, if the desired change in μ is present, it should be evident by about the $r = 10$th failure and the test can be terminated early. Thus, the potential for time savings and component savings are evident. Interestingly, one of our goals was to find the exact power functions instead of using simulation to compute power. Due to the complexity of sending data from one distribution thru the PIT of another, the resulting transformations were so complicated that we could only find the exact power function for the exponential parent with $r = 2$. We are currently working on parameter estimation using the inverse test statistic technique on T_r. We are establishing confidence intervals for one-parameter distributions and confidence regions for two or more parameter distributions. Research can also be extended to the case where the null hypothesis distribution is not fully specified, but must be estimated from the data, a much more difficult case.

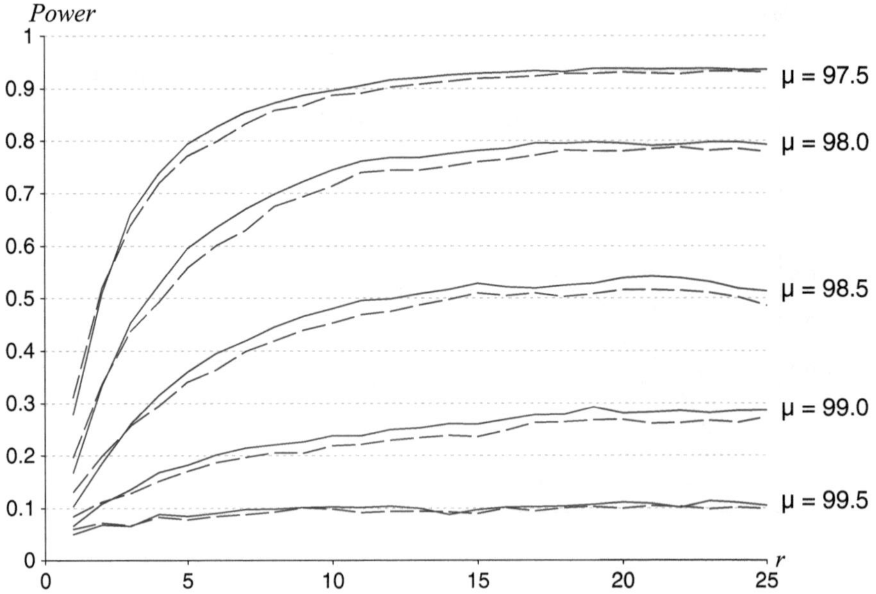

Fig. 6.3. Power curves displaying the results of Monte Carlo power simulation with underlying gamma distribution of the data, $\sigma^2 = 10$, $\alpha = 0.05$ and $n = 25$. Under H_0, $\mu = 100$. The *solid line* is the T_r power curve and the *dashed line* is the A–D power curve. Note, as the true value of μ decreases from 100, the test based on the T_r test statistic outperforms the test based on the A–D test statistic

Appendix

In this appendix, we will show the computations needed to find the p-value for a small censored sample. Assume there exists a mechanical system that has an established time-to-completion that is modeled by the gamma(2.1, 4.41) distribution, where time is measured in hours. A new process is developed and experimenters hope to show an improvement (decrease) in completion time. The new process is applied simultaneously with $n = 25$ different machines and it is noted that the first five completion times are 0.40, 0.54, 0.66, 0.75, and 0.77 h. Completion of the full experiment, under the null hypothesis, has an expected time of $E(X_{(25)}) \cong 4.52$ h, the expected completion time of the slowest machine. However, the fifth machine to finish has an expected completion time, under the null hypothesis, of $E(X_{(5)}) \cong 1.21$ h. Because the observed time of the fifth machine's completion was only 0.77 h, it would be useful to know if there is enough statistical evidence to stop the experiment, concluding that the new process is faster. The following APPL code will analyze this Type II censored experiment.

```
> Old_Process := GammaRV(2.1, 4.41);
> n := 25;
> data := [0.40, 0.54, 0.66, 0.75, 0.77];
> CensoredT(Old_Process, data, n);
```

The procedure output is the test statistic, the lower tail p-value, and the upper tail p-value. In this case, those values are 1.038165719, 0.0100496, 0.989950. Because we consider this a two-tailed test, we have a p-value of $0.0100496 \times 2 = 0.02$, significant evidence that the new process is faster and we can consider terminating the experiment.

The calculations that follow are intended to explain more fully how the process works. The existing process has time to completion distribution X distributed according to the gamma(2.1, 4.41) random variable with PDF $f(x) = 2.565595x^{3.41}e^{-2.1x}$, $0 < x < \infty$, and CDF, calculated in APPL, of

$$\begin{aligned}F(x) = {} & \frac{100000}{66987458901\Gamma(41/100)} x^{\frac{41}{200}} 21^{\frac{41}{200}} 10^{\frac{159}{200}} e^{-\frac{21}{20}x} \Big[82181\texttt{WhittakerM}(\frac{41}{200}, \frac{141}{200}, \frac{21}{10}x) \\ & +143220\texttt{WhittakerM}(\frac{41}{200}, \frac{141}{200}, \frac{21}{10}x)x + 132300\texttt{WhittakerM}(\frac{41}{200}, \frac{141}{200}, \frac{21}{10}x)x^2 \\ & -71610x\texttt{WhittakerM}(-\frac{159}{200}, \frac{141}{200}, \frac{21}{10}x) - 82181\texttt{WhittakerM}(-\frac{159}{200}, \frac{141}{200}, \frac{21}{10}x) \\ & +92610x^3\texttt{WhittakerM}(\frac{41}{200}, \frac{141}{200}, \frac{21}{10}x) - 44100x^2\texttt{WhittakerM}(-\frac{159}{200}, \frac{141}{200}, \frac{21}{10}x)\Big],\end{aligned}$$

$x > 0$, relying on Maple's `WhittakerM` function, a solution to a differential equation. The null hypothesis is that $\mu_{\text{new}} = \mu_{\text{old}}$ and the alternate hypothesis is that $\mu_{\text{new}} \neq \mu_{\text{old}}$. The algorithm first transforms these data values thru the gamma CDF, $z_{(i)} = F(x_{(i)})$, $i = 1, 2, \ldots, 5$ (the first PIT) to come up with five transformed values 0.005203, 0.015494, 0.030823, 0.046777, and 0.050857.

For each of the $z_{(i)}$, it is necessary to calculate the appropriate conditional order statistic CDF so that the un-ordered uniform variates can be calculated. The first data point has CDF

$$\begin{aligned}F_{X_{(1)}}(x) = {} & x^{25} - 25x^{24} + 300x^{23} - 2300x^{22} + 12650x^{21} - 53130x^{20} + 177100x^{19} \\ & -480700x^{18} + 1081575x^{17} - 2042975x^{16} + 3268760x^{15} - 4457400x^{14} \\ & +5200300x^{13} - 5200300x^{12} + 4457400x^{11} - 3268760x^{10} + 2042975x^{9} \\ & -1081575x^{8} + 480700x^{7} - 177100x^{6} + 53130x^{5} - 12650x^{4} + 2300x^{3} \\ & -300x^{2} + 25x,\end{aligned}$$

for $0 < x < 1$. Thus, the first iid $U(0, 1)$ p-value (from the second PIT) is $F_{X_{(1)}}(0.005203) = 0.122263$. The second data value comes from the truncated order statistic from the U(0, 1) distribution with $n = 24$ and $r = 1$ truncated on the left at $z_{(1)} = 0.005203$. The CDF for this point is

$$
\begin{aligned}
F(x) = & -1.1334x^{24} + 27.2008x^{23} - 312.809x^{22} + 2293.94x^{21} - 12043.2x^{20} \\
& +48172.6x^{19} - 152546.7x^{18} + 392262.9x^{17} - 833558.6x^{16} + 1481882.0x^{15} \\
& -2222823.1x^{14} + 2829047.5x^{13} - 3064801.5x^{12} + 2829047.5x^{11} - 2222823.0x^{10} \\
& +1481882.0x^{9} - 833558.6x^{8} + 392262.9x^{7} - 152546.7x^{6} + 48172.6x^{5} \\
& -12043.2x^{4} + 2293.94x^{3} - 312.809x^{2} + 27.201x - 0.1334,
\end{aligned}
$$

for $0.005203 < x < 1$. The second p-value is $F(0.015494) = 0.220857$. Similarly, the third data value comes from the truncated order statistic U(0, 1) distribution from $n = 23$ and $r = 1$, truncated on the left at $z_{(2)} = 0.01549350642$. The CDF for this point is

$$
\begin{aligned}
F(x) = & \ 1.4320x^{23} - 32.9382x^{22} + 362.321x^{21} - 2536.24x^{20} + 12681.2x^{19} \\
& -48188.6x^{18} + 144565.9x^{17} - 351088.7x^{16} + 702177.4x^{15} - 1170295.6x^{14} \\
& +1638413.9x^{13} - 1936307.3x^{12} + 1936307.3x^{11} - 1638413.9x^{10} + 1170295.6x^{9} \\
& -702177.4x^{8} + 351088.7x^{7} - 144565.9x^{6} + 48188.6x^{5} - 12681.2x^{4} \\
& +2536.24x^{3} - 362.32x^{2} + 32.94x - 0.4321,
\end{aligned}
$$

for $0.015494 < x < 1$. The third p-value is $F(0.030823) = 0.302984$. Likewise, the fourth data value comes from CDF

$$
\begin{aligned}
F(x) = & -1.9913x^{22} + 43.8082x^{21} - 459.986x^{20} + 3066.58x^{19} - 14566.2x^{18} \\
& +52438.4x^{17} - 148575.6x^{16} + 339601.3x^{15} - 636752.4x^{14} + 990503.7x^{13} \\
& -1287654.9x^{12} + 1404714.4x^{11} - 1287654.9x^{10} + 990503.7x^{9} - 636752.5x^{8} \\
& +339601.3x^{7} - 148575.6x^{6} + 52438.4x^{5} - 14566.2x^{4} + 3066.58x^{3} \\
& -459.986x^{2} + 43.808x - 0.99128,
\end{aligned}
$$

for $0.030823 < x < 1$. The fourth p-value is $F(0.046777) = 0.305921$. Finally, the fifth data value comes from the CDF

$$
\begin{aligned}
F(x) = & \ 2.7348x^{21} - 57.4298x^{20} + 574.298x^{19} - 3637.22x^{18} + 16367.5x^{17} \\
& -55649.5x^{16} + 148398.7x^{15} - 317997.2x^{14} + 556495.0x^{13} - 803826.2x^{12} \\
& +964591.4x^{11} - 964591.4x^{10} + 803826.2x^{9} - 556495.0x^{8} + 317997.2x^{7} \\
& -148398.7x^{6} + 55649.5x^{5} - 16367.5x^{4} + 3637.22x^{3} - 574.298x^{2} \\
& +57.4298x - 1.7347,
\end{aligned}
$$

for $0.046777 < x < 1$. The fifth p-value is $F(0.050857) = 0.086139$. The test statistic is the sum of the five p-values, $t = 1.038166$, and has a CDF based on the null hypothesis of that of convolution of five U(0, 1) random variables:

$$F(x) = \begin{cases} 0 & x < 0 \\ \frac{1}{120}x^5 & 0 < x < 1 \\ \frac{1}{24} + \frac{5}{12}x^2 - \frac{5}{12}x^3 + \frac{5}{24}x^4 - \frac{1}{30}x^5 - \frac{5}{24}x & 1 < x < 2 \\ -\frac{21}{8} + \frac{155}{24}x - \frac{25}{4}x^2 + \frac{35}{12}x^3 - \frac{5}{8}x^4 + \frac{1}{20}x^5 & 2 < x < 3 \\ \frac{141}{8} - \frac{655}{24}x + \frac{65}{4}x^2 - \frac{55}{12}x^3 + \frac{5}{8}x^4 - \frac{1}{30}x^5 & 3 < x < 4 \\ -\frac{601}{24} + \frac{625}{24}x - \frac{125}{12}x^2 + \frac{25}{12}x^3 - \frac{5}{24}x^4 + \frac{1}{120}x^5 & 4 < x < 5 \\ 1 & 5 < x < \infty\,. \end{cases}$$

The lower tail p-value is therefore $2 \times F(1.309743407) = 2 \times 0.01000496 \cong 0.0200$, a low p-value suggesting the expected completion time for the new process is faster than that of the existing process.

7

Maximum Likelihood Estimation Using Probability Density Functions of Order Statistics

Andrew G. Glen

Abstract A variation of maximum likelihood estimation (MLE) of parameters that uses PDFs of order statistic is presented. Results of this method are compared with traditional maximum likelihood estimation for complete and right-censored samples in a life test. Further, while the concept can be applied to most types of censored data sets, results are presented in the case of order statistic interval censoring, in which even a few order statistics estimate well, compared to estimates from complete and right-censored samples. Population distributions investigated include the exponential, Rayleigh, and normal distributions. Computation methods using APPL are simpler than existing methods using various numerical method algorithms.

Keywords Computational probability • Interval censoring • Life tests

7.1 Introduction

Using the PDFs of order statistics has some intuitive appeal when estimating parameters for survival distributions. In a life test, for example, n items are placed on test simultaneously and run until failure. As items fail, the failure

---

Originally published in *Computers and Industrial Engineering*, Volume 58, Issue 4, in 2010, this paper relied extensively on the APPL environment to explore and analyze censored data techniques. At the heart of the research was the need to find likelihood functions for various censoring schemes. These functions needed the PDFs of order statistics, and the APPL `OrderStat` procedure produced them. The simulated results reported in the last table all came from APPL-based simulations.

A.G. Glen (✉)
Department of Mathematics and Computer Science, The Colorado College, Colorado Springs, CO, USA
e-mail: andrew.glen.1984@gmail.com

A.G. Glen, L.M. Leemis (eds.), *Computational Probability Applications*, International Series in Operations Research & Management Science 247,
DOI 10.1007/978-3-319-43317-2_7

times are presented in order, from the smallest failure time to the largest failure time. These ordered failure times are in fact order statistics. Life tests have the advantage of presenting data as order statistics, so the properties of order statistics can be used to analyze the experiment, even before the nth item fails. Oftentimes some sort of censoring of the data occurs, the most common being with only the first r failure times are known, a case called Type II right-censoring.

Let n items from a population with lifetime modeled by a random variable X have a life distribution with PDF f_X. Let $X_{(1)}, X_{(2)}, \ldots, X_{(n)}$ be the order statistics if the experiment was allowed to continue until all n items failed. Each of these statistics has a PDF, as introduced in mathematical statistics books. If all n items fail, then a typical likelihood function takes on the form

$$L(\theta) = \prod_{i=1}^{n} f_X(x_i, \theta).$$

If only the first r items are observed to fail, then it is typical to use the resulting Type II right-censored likelihood function that is proportional to

$$L(\theta) = \prod_{i=1}^{r} f_X(x_{(i)}, \theta) \times \left(1 - F_X(x_{(r)}, \theta)\right)^{n-r},$$

in which F_X is the CDF of X. But if random censoring occurs, where only some order statistics are known, this article proposes constructing likelihood equations based on the PDFs of the distributions of only the known order statistics. In other words, let K be the set of order statistic indices for the observed failure times, then one could construct the likelihood function

$$L_K(\theta) = \prod_{i \in K} f_{X_{(i)}}(x_{(i)}, \theta),$$

in which $f_{X_{(i)}}$ is the PDF of the marginal distribution of the ith order statistic. This method will be referred to as maximum likelihood estimation with order statistics (MLEOS). In Section 7.3, we will show the utility of this type of estimation.

The literature on estimating with order statistics is very comprehensive. However, in almost every article or book, order statistics are used in closed form estimation formulas, based on certain properties that some distributions might have as advantages. For example, Leemis and Shih [97] derive estimators for the parameters of the exponential and Rayleigh distributions for left-censored data sets. In his comprehensive study on order statistics, David and Nagaraja [43, p. 245] devotes all of the sixth chapter to a discussion on estimating and hypothesis testing with order statistics. All of his results, though, are again limited to closed-form estimators that use the values of the order statistics, not the PDFs of the order statistics. Estimators that can be expressed in closed form often have advantages over those that require numerical methods for their calculation. Some of the estimators that can be expressed in closed form allow the calculation of the distribution of the estimator, which can be used, for example, to derive exact confidence intervals. However, some of these closed-form estimators are dependent on underlying distributions for some of these advantages. This method's proposal of a non-closed form estimator can be applied to many distributions with many types of parameters. Usually, non-closed-form estimators require some form of numerical method to calculate the estimates, Newton–Raphson or expectation–maximization (EM) algorithms for examples. This research uses of APPL [46] to make direct use of calculus on exact likelihood functions to estimate parameters. Further, APPL, an open source set of procedures available from the author, is specifically designed to use PDFs and create new ones where necessary. APPL in effect turns Maple into a mathematical statistics pallet for designing new distributions as well as simulating more complicated random functions without resorting to heavy coding in more basic computer languages. Thus, some estimates are found exactly, without numerical methods. In the cases where numerical solutions are needed, it is a more seamless action using built-in commands of the computer algebra software. A side advantage of using APPL is that one's knowledge of estimating parameters does not necessarily need to be augmented with knowledge of numerical methods.

7.2 MLEOS with Complete Samples

First, MLEOS is presented in the case of complete samples. It is not expected to improve upon standard MLE for complete samples, since the standard methods are asymptotically minimum variance and unbiased. However, it is instructive to see the intuition of the process, and this intuition is easily shown below with complete samples. The MLEOS approach to estimation borrows from the intuition associated with traditional maximum likelihood estimation. In his Figure 7.7, Leemis [93, pp. 172–173] gives a graphical interpretation of the MLE method. Leemis suggests the goal of MLE is to find the parameter θ that "maximizes the product of the density values at the data points," which he shows graphically using a number of vertical lines underneath one exponential PDF. Since in MLE one seeks a value of θ that maximizes the product of these density values from the PDF, it is a logical extension to investigate what would happen if the product of the density values of the n order statistics was maximized. Maximizing the product of the "hump-shaped" order statistic PDFs evaluated at their respective ordered data values is graphically depicted in Figure 7.1, the PDFs of the four order statistics of the unit exponential distribution with $n = 4$. In effect one is finding a $\hat{\theta}$ value that maximizes the product of the lengths of the vertical lines (the lines from the ordered data points to their ordered density values). A good estimate for θ 'moves' the centers of the n humps of the n PDFs near the data values. Therefore, MLEOS finds a $\hat{\theta}$ that maximizes the products of the respective density values. While

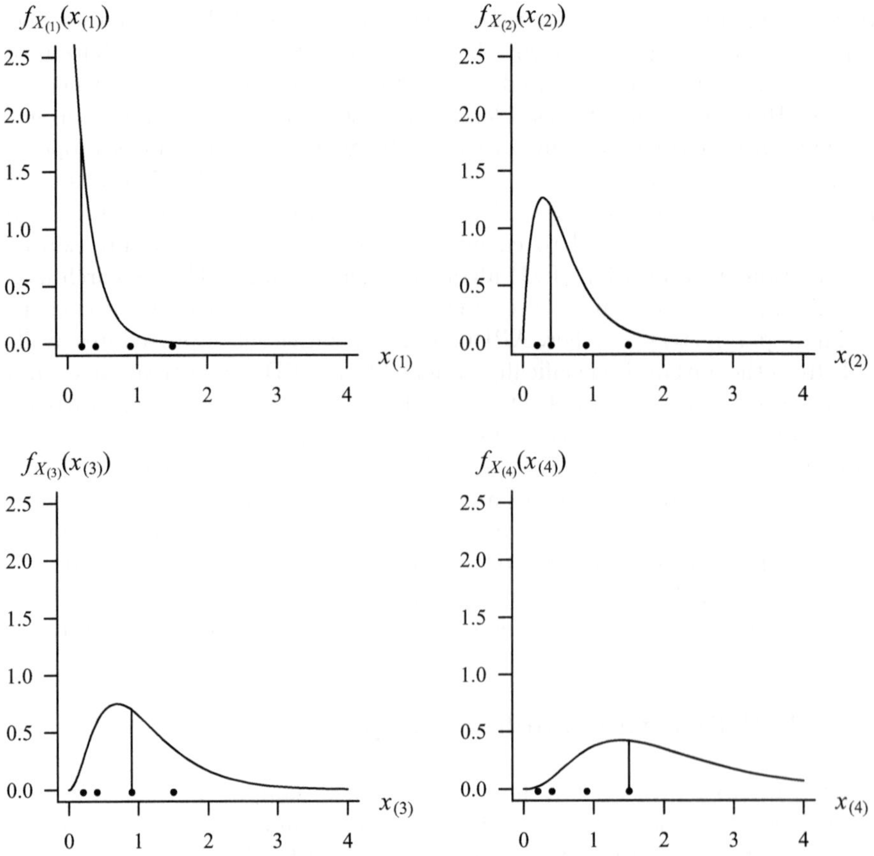

Fig. 7.1. The PDFs of the four order statistics from a complete sample of $n = 4$ from a unit exponential distribution. The *vertical lines* represent the four factors that comprise the likelihood function for MLEOS

MLE has the property that the likelihood function is also the joint distribution PDF, this is not the case for MLEOS. Since the order statistics are dependent, the product for L_K is not a joint distribution function. Note, in Figure 7.1, the filled circles on the horizontal axis correspond to the complete sample [0.2, 0.4, 0.9, 1.5]. For the case of complete data sets, parameter estimates were computed from various population distributions from simulation using both MLE and MLEOS. The variance of the two estimation techniques were compared. Complete sample estimates for θ from the exponential distribution, μ and σ from the normal distribution, and θ from the $U(0, \theta)$ distribution were investigated. Of all these population distributions, only the last case, estimating θ from the uniform distribution, resulted in a slight improvement in the mean of the estimator. Since some maximum likelihood (ML) estimators are also minimum variance unbiased estimators (MVUE), it is not expected that many MLEOS estimates will outperform MLE with respect to variance in the case of complete samples with sufficient sample size. This is the case for estimating θ for the exponential distribution and μ for the normal distribution [77]. However, the parameter σ from the normal distribution is not a MVUE [77, p. 10] and the MLE for θ in the uniform distribution is biased [87, p. 244]. For the uniform distribution, the reason that the MLEOS estimator was less biased than the MLE estimator is easy to see. Each MLE $\hat{\theta}$ is the largest data value in the sample. The MLEOS $\hat{\theta}$ is either the maximum data value, or the solution to $\frac{\partial L(\mathbf{x},\theta)}{\partial \theta} = 0$. While it is possible that the maximum likelihood value of $\hat{\theta}$ is less than the maximum of the sample, that maximum is clearly a lower bound of the parameter. In about 54 % of the samples, the solution to this equation was greater than the maximum of the sample. The comparisons of the MLE, an unbiased estimator, and the MLEOS estimator are given in Table 7.1. Analytic methods were used to calculate $E(\hat{\theta})$ and $V(\hat{\theta})$ for MLE and unbiased estimate. Simulation was used for approximating $E(\hat{\theta})$ and $V(\hat{\theta})$ for MLEOS. All results in Table 7.1 were based on a sample size of $n = 25$. The simulation was based on 50,000 replication. It was assumed that $\theta = 1$. The MLEOS estimator had less bias but higher variance than the MLE estimator. As could be expected, however, the unbiased estimator is better for both bias and variance than MLEOS.

While applying MLEOS to complete samples is interesting and instructive for understanding the concept, it is generally not expected to outperform standard MLE, as that method is well established and has many useful properties. The real advantage to MLEOS becomes apparent in censored samples, as presented in the next section.

7.3 Applying MLEOS to Censored Samples

An important area that benefits from MLEOS is found in life tests with censoring. Censoring mechanisms include left-censoring, right-censoring of Type I and Type II, and random censoring. Helpful explanations of these types of censoring are presented in Leemis [93, p. 184] and Lee [92, pp. 2–3], and especially in the graphs on page 8 of Nelson [119]. Further information on the subject

Table 7.1. The variance of the MLE, MLEOS, and MLE with unbiasing factor for the uniform distribution on a complete sample of $n = 25$ observations

Technique	$E(\hat{\theta})$	$V(\hat{\theta})$
MLE (exact)	$25/26 \cong 0.9615$	$25/18252 \cong 0.001370$
MLEOS (simulated)	0.9882	0.003018
MLE with unbiasing factor (exact)	1	$1/675 \simeq 0.001481$

of estimation is available from Lawless [91], Deshpande and Purohit [45] and Kalbfleisch and Prentice [75]. Klein and Moeschberger [79, Chap. 3] describe the more general cases of doubly-censored data, progressively censored data, and interval censored data. They describe a general likelihood function that encompasses in part the occurrence of interval censoring. While much is reported in the literature for right-censoring, less is available on left-censoring, and even less for interval censoring. Oller et al. [124] considered likelihood estimation in the case when the random nature of the interval is assumed to be set in advance. This paper's proposed estimation with MLEOS is not limited to that assumption. Odell et al. [123] compared maximum likelihood estimation with and without an assumed midpoint in an interval-censored data simulation for an accelerated failure time regression model, using numerical methods to solve for the estimates, and a relatively small number of replications (100). Estimation with MLEOS does not need an assumed midpoint and employs the computer algebra combination of Maple and APPL to simulate more replications (10,000) and compute exact estimates (in some cases). Sun [157] discusses non-parametric estimation of survival distributions with a separate (but similar) type of censoring, random censoring, however there is a clear distinction between interval censoring and random censoring, pointed out by a number of these references.

Interval censoring occurs in a life test when a subject's lifetime is not known, but left and right limits of the lifetime are known for that subject. Sometimes in interval censoring, while the lower and upper limits of a failure time are known, there may be no single failure time observed. An example of such interval censoring is given on page 8 of Nelson [119]. This paper considers a special case of interval censoring that will be termed "order statistic interval censoring." Such is the case when an item's failure time is known, or its unknown failure time is before the first known order statistic, between two known order statistics, or after the last known order statistic. Consider, for example, 30 subjects on a life test. By time L, four subjects have failed (and are observed), at times $x_{(1)}$, $x_{(2)}$, $x_{(3)}$, $x_{(4)}$. The observers then leave for the day. By time R, the observers return to find three additional subjects failed overnight, censoring the values of the 5th, 6th, and 7th order statistics. Then the 8th failure is observed, $x_{(8)}$, upon which the life test is terminated similarly to a Type II right-censored experiment. In this case the three missing observations are in the interval (L, R) but they are also bounded by $x_{(4)}$ and $x_{(8)}$. Since only the first four and the 8th order statistics are known, the likelihood function is comprised of these five order statistics PDFs defined by the set $K = \{1, 2, 3, 4, 8\}$:

$$L_K(\theta) = f_{X_{(1)}}(x_{(1)}, \theta) \times \ldots \times f_{X_{(4)}}(x_{(4)}, \theta) \times f_{X_{(8)}}(x_{(8)}, \theta).$$

In general, the MLEOS likelihood function for order statistic interval censoring is as follows:

$$L_K(\theta) = \prod_{i \in K} f_{X_{(i)}}(x_{(i)}, \theta),$$

where K is the set of order statistic indices for the observed failure times. Note, it is not necessary to add survivor functions to the likelihood function, as in other interval censoring techniques, e.g., Klein and Moeschberger [79, p. 66], since the PDFs of the existing order statistics help account for the missing information from the censored items. Clearly order statistic interval censoring is a more general censoring mechanism that includes all Type II right-censoring as well as some left and doubly censoring schemes.

It is generally preferred to find a closed-form function for parameter estimation. In other words one wants to find a function for $\hat{\theta}$ that is only a function of the known data, i.e., $\hat{\theta} = f(X_{(i \in k)})$, which is the case with many estimators with complete samples. Unfortunately, with order statistic PDFs, only part of this desired condition is attainable. Consider the case of knowing only two order statistics under assumed exponential data with the PDF

$$f(x) = \frac{1}{\theta} e^{-x/\theta} \qquad x, \theta > 0.$$

Let the two known order statistics be $x_{(l)}$ and $x_{(r)}$ with $0 < l < r \leq n$ and $K = \{l, r\}$. The maximum likelihood function with order statistic interval censored data is therefore proportional to

$$L_K(\theta) = f_{X_{(l)}}(x_{(l)}) \cdot f_{X_{(r)}}(x_{(r)})$$

where $f_{X_{(l)}}$ is the PDF of the marginal distribution for the lth order statistic and $f_{X_{(r)}}$ is the PDF of the marginal distribution for the rth order statistic. Substituting in the PDFs of the exponential order statistics, the likelihood function becomes

$$L_K(\theta) = \frac{(n!)^2 \left(1 - e^{-x_{(l)}/\theta}\right)^{l-1} \left(e^{-x_{(l)}/\theta}\right)^{n-l} e^{-x_{(l)}/\theta} \left(1 - e^{-x_{(r)}/\theta}\right)^{r-1}}{(l-1)!(r-1)!(n-l)!(n-r)!\theta^2} \cdot \left(e^{-x_{(r)}/\theta}\right)^{n-r} e^{-x_{(r)}/\theta}.$$

The log likelihood function is therefore

$$\ln L_K(\theta) = -2 \ln \theta + \ln A,$$

where

$$A = \frac{(n!)^2 \left(1 - e^{-x_{(l)}/\theta}\right)^{l-1} e^{-\frac{x_{(l)}n - x_{(l)}l + x_{(l)} + x_{(r)}n - x_{(r)}r + x_{(r)}}{\theta}} \left(1 - e^{-x_{(r)}/\theta}\right)^{r-1}}{(l-1)!(r-1)!(n-l)!(n-r)!}.$$

The derivative of $\ln L_K$ is

$$\frac{\partial \ln L_K(\theta)}{\partial \theta} = \frac{-B}{\theta^2 \left(-1 + e^{-x_{(l)}/\theta}\right)\left(-1 + e^{-x_{(r)}/\theta}\right)},$$

where

$$\begin{aligned}
B = {} & 2\theta - 2\theta e^{-x_{(r)}/\theta} - 2\theta e^{-x_{(l)}/\theta} + 2\theta e^{-\frac{x_{(l)}+x_{(r)}}{\theta}} + x_{(r)} e^{-x_{(l)}/\theta} + x_{(l)} e^{-x_{(r)}/\theta} \\
& - x_{(l)} n e^{-\frac{x_{(l)}+x_{(r)}}{\theta}} - x_{(r)} n e^{-\frac{x_{(l)}+x_{(r)}}{\theta}} + x_{(l)} n e^{-x_{(l)}/\theta} - x_{(l)} l e^{-x_{(r)}/\theta} \\
& + x_{(l)} n e^{-x_{(r)}/\theta} - x_{(r)} r e^{-x_{(l)}/\theta} + x_{(r)} n e^{-x_{(l)}/\theta} + x_{(r)} n e^{-x_{(r)}/\theta} \\
& - x_{(l)} - x_{(r)} - x_{(l)} n + x_{(l)} l - x_{(r)} n + x_{(r)} r.
\end{aligned}$$

Setting the first derivative equal to zero and solving for θ results in the following equation for the estimator

$$\hat{\theta} = \frac{-x_{(r)}}{\texttt{RootOf}(C)},$$

where

$$\begin{aligned}
C = {} & 2x_{(r)} e^{z + z x_{(l)}/x_{(r)}} - 2e^{z} x_{(r)} - e^{z} z n x_{(r)} - e^{z} z x_{(l)} + z x_{(l)} n e^{z + z x_{(l)}/x_{(r)}} \\
& + z n x_{(r)} e^{z + z x_{(l)}/x_{(r)}} + e^{z} z x_{(l)} l - e^{z} z x_{(l)} n - 2e^{z x_{(l)}/x_{(r)}} x_{(r)} + 2x_{(r)} + z x_{(l)} \\
& + z x_{(r)} + z x_{(l)} n - z e^{z x_{(l)}/x_{(r)}} x_{(r)} - z x_{(l)} n e^{z x_{(l)}/x_{(r)}} + z r e^{\frac{z x_{(l)}}{x_{(r)}}} x_{(r)} \\
& - z n e^{z x_{(l)}/x_{(r)}} x_{(r)} - z x_{(l)} l + z n x_{(r)} - z r x_{(r)},
\end{aligned}$$

and the `RootOf` command is a Maple procedure that evaluates the root of the expression for variable z. This expression was found with APPL running inside a Maple session. It is possible to isolate the parameter θ and find an expression for the value of $\hat{\theta}$ that maximizes the likelihood function. Unfortunately, while the expression for $\hat{\theta}$ is calculable, finding the distribution of the estimator is not. The transformation implied in this equation is far too complicated for transformation techniques available in mathematical statistics. Thus, expressions for bias, variance and confidence intervals of the estimator can not be found analytically, but must be simulated. Of course, as more order statistics are known, and the population PDFs become more complicated, these likelihood functions become even more untenable. Thus, as is typical with MLE under censorship of various types, one must resort to simulation to calculate properties of the estimates, such as bias, variance, and mean square error of the estimates.

To show the value of MLEOS in order statistic interval censoring, a set of simulations was conducted for various censoring patterns. Data generated from the exponential, Rayleigh, and normal distributions were then used to estimate parameters for the complete sample with $n = 30$, Type II right-censored sample and different patterns of known order statistics. Comparisons are made between the best case of all r order statistics observed (estimated by MLE) versus degraded cases where only a subset up to and including the rth order statistic are observed (estimated by MLEOS). The results of these simulations are shown in Table 7.2.

In each case, a sample of 30 variates was created, then the different estimates were calculated from that sample so that a comparison of the average and the variance of each estimator was possible. Mean square error of the estimate was also considered and behaved similarly to estimator variance, thus, while available from the author, is omitted from discussion here. There were 10,000 samples created for the exponential and Rayleigh simulations and 2500 samples created for the normal distribution. These simulations were conducted with APPL running in a Maple session. The code is available from the editors. The exponential and Rayleigh distributions were chosen because they are well known, one-parameter survival distributions. Parameterizations for these two distributions are, respectively,

$$f(x) = \frac{1}{\theta} e^{-x/\theta} \qquad x > 0, \text{ and}$$

$$f(x) = \frac{2x}{\theta^2} e^{-x^2/\theta^2} \qquad x > 0.$$

The normal distribution, while not typically a survival distribution, was chosen because of its infinite tail on the left of the support, resulting in an unexpected outcome that is mentioned in the next paragraph. For the normal distribution, the mean was fixed at $\mu = 100$ (arbitrarily) and the standard

Table 7.2. Outcomes of simulations for various estimates and various distributions The darkened circles represent the order statistics that were observed, while the clear circles represent the order statistics that were not observed. Even though up to $r = 20$ order statistics are shown, the data are censored from a set with $n = 30$

	Normal, σ = 10		Rayleigh θ = 100		Exponential θ = 100	
Type of estimation	**average**	**variance**	**average**	**variance**	**average**	**variance**
MLE complete sample n = 30	9.97	1.73	99.53	83.52	99.94	328.23
MLEOS ○●●○○○○○○○○○○○○○○○○○	9.82	4.80	96.04	859.15	100.11	3595.52
MLE ●●●○○○○○○○○○○○○○○○○○	9.78	3.83	95.95	770.77	102.69	3573.50
MLEOS ●○●○●○○○○○○○○○○○○○○○	9.83	4.13	97.51	552.03	99.84	2282.39
MLE ●●●●●○○○○○○○○○○○○○○○	9.74	3.64	97.40	452.81	101.80	1989.89
MLEOS ○●○●○●○●○○○○○○○○○○○○	9.69	4.76	98.39	351.54	99.96	1450.03
MLE ●●●●●●●●○○○○○○○○○○○○	9.75	3.58	98.06	292.52	100.75	1203.56
MLEOS ●●●●●●●●●●○○○○○○○○○○	9.69	4.55	98.67	295.16	99.93	1210.62
MLE ●●●●●●●●●●○○○○○○○○○○	9.77	3.51	98.54	241.11	100.93	989.57
MLEOS ○○○○○○○○○○○●●○○○○○○○	8.46	24.12	99.16	191.00	100.00	790.07
MLE ●●●●●●●●●●●●●○○○○○○○	9.80	3.41	98.94	184.42	100.62	736.19
MLEOS ○○○○○○○●○●○●○●○○○○○○	9.09	11.61	99.15	187.52	99.95	769.93
MLE ●●●●●●●●●●●●●●○○○○○○	9.81	3.21	99.00	171.79	100.46	688.18
MLEOS ○○○○○○○○○○●○●○●○○○○○	8.83	18.97	99.24	171.16	100.07	700.42
MLEOS ○○○○○●●●●●●●●●●○○○○○	9.28	8.69	99.17	184.30	99.98	756.86
MLE ●●●●●●●●●●●●●●●○○○○○	9.81	3.13	99.09	161.33	100.46	644.68
MLEOS ○○○○○○○○○○○○○○○○○○●●	8.45	19.98	99.38	133.85	100.52	543.75
MLEOS ○○○○○○○○○○○○○○○●○●○●	8.81	18.22	99.41	131.85	100.49	534.64
MLEOS ○○○○○○○○○○○○○●○●○●○●	9.11	16.23	99.40	131.69	100.39	534.08
MLEOS ○○○○○○○○○○●●●●●●●●●●	9.41	13.16	99.39	135.28	100.26	547.27
MLEOS ●●●●●●●●●●●●●●●●●●●●	9.75	3.78	99.35	144.85	100.14	587.53
MLE ●●●●●●●●●●●●●●●●●●●●	9.86	2.55	99.28	123.29	100.06	499.57

deviation was estimated from the sample, similar to the process of finding the distribution of residuals in regression. In Table 7.2 the status of the order statistics is indicated the clear circles for censored times and dark circles for observed times. Each MLEOS estimate is compared with the associated MLE method for all r observations. For example, in the first MLEOS estimate, the 2nd and 3rd circles are darkened, indicating that only the 2nd and 3rd order statistics were observed. This is compared in the next line with an MLE estimate based on all $r = 3$ observations known. Note how in the case of the exponential distribution, the estimates have identical mean and variance (between the MLEOS and the MLE estimates). For both the exponential and Rayleigh distributions, the MLEOS estimates with the later lifetimes observed (e.g., closest to the 20th order statistic), gave the lowest variance and least evident bias in the mean. Even in the case where only the 19th and 20th order statistics were known, the mean and variance were very close to that of the right-censored MLE. This is remarkable, because in effect observing the first 18 failure times gave little additional accuracy to $\hat{\theta}$ when $r = 20$. Also, for the exponential and Rayleigh distributions, the mean of the estimators with later order statistics is closer to the true mean than the Type II right-censored estimator. This too is remarkable, since the standard MLE estimator uses more failure times, but produces a more biased estimate. For these two distributions, the MLEOS that only had the earliest order statistics known gave the highest variance and substantial evidence of bias. Interestingly, the opposite is true for the normal distribution. The first few order statistics by themselves gave almost the same quality of estimate as the complete and right-censored estimates. These estimators contain as few as two order statistics, the 2nd and 3rd. To achieve that level of quality in an estimator with only a few early observed order statistics is very useful. Conversely, the estimates comprised of only the later observed order statistics suffered from high variance and apparent bias. The last two rows of estimates calculate $\hat{\theta}$ when all $r = 20$ order statistics are observed for both MLEOS and MLE. This is for comparison only as standard MLE for Type II right censoring should be used as that method has many statistical advantages that are well established in the literature.

In some distributions, the early order statistics give better estimates, and in other distributions the later order statistics are better. Recall that the early order statistics of the exponential distribution are high in variance, thus, poorer estimates. The opposite is true for the early order statistics of the normal distribution, which have low variance, possibly due to the support of negative infinity on the left. A possible explanation for the exponential distribution's behavior can be found in the fourth graph of Figure 7.1. That fourth order statistic distribution has a high variance, thus, the value of $x_{(4)}$ has a lot of room to move and still not effect the value of $\hat{\theta}$ by much. Conversely, the first order statistic graphed shows that small movements in $x_{(1)}$ can have a disproportionately large effect on the value of $\hat{\theta}$.

Clearly these results have some implications about conducting life tests. In the case in which missing data points exist, this study provides another method for inference. More deliberately, though, this study could be used as justification to reduce resource requirements of observing life tests. Depending on expected failure rates, researchers can relax to some extent the vigilance needed to observe certain types of life tests, especially those that are expensive to have complete time observers. Also, some medical life tests with only periodic checking of subjects may benefit from this type of inference.

7.4 Conclusions and Further Research

This paper presents an estimation technique that uses PDFs of order statistics to create maximum likelihood estimates. The method can be applied to many types of complete and censored samples, especially in the area of life tests. The method can be used for any continuous survival distribution. Promising results are presented in the case of interval censoring of life test data, where the intervals are defined by the known order statistics, for the case of the exponential, Rayleigh, and normal distributions. Tabular results of estimation simulations show how MLEOS estimates are almost as good as Type II right-censored estimates, even in the case of only observing two order statistics. Implications are presented for the design of life tests, especially in the conduct of the test itself.

8

Notes on Rank Statistics

Daniel J. Luckett, Samatha King, and Lawrence M. Leemis

Abstract We consider certain aspects of two ranks statistics. For the non parametric Wilcoxon signed-rank test, power curves for this test are plotted and compared for several population distributions. The power of the Wilcoxon signed-rank test is compared to the power of the sign test. Computations to calculate power are performed using a computer algebra system and an algorithm is presented to perform these computations. Power curves are plotted and compared for small sample sizes. Monte Carlo simulation is used to calculate power curves for larger sample sizes. For the Mann–Whitney rank sum statistic, the distribution of the statistic is considered in the presence of ties, both within a sample and between samples.

Keywords Hypothesis testing • Non parametric tests • Power

This original paper presents research using APPL as the computing environment to explore special situations of two rank statistics. An interesting application using APPL here is creating power curves. Power curves for different distributions are typically not part of an statistical package. This paper was able to use the survival function ability and a few other APPL functions to create exact power curves for different sample sizes for the Wilcoxon test. Some of the discrete processes in APPL can be used also for the Mann–Whitney rank test to find exact distributions, especially in the case of ties in the data. Of interest is the new APPL procedure `WMWRV` which calculates the exact distribution of H_0 without ties.

D.J. Luckett • S. King • L.M. Leemis (✉)
The College of William and Mary, Williamsburg, VA, USA
e-mail: leemis@math.wm.edu

A.G. Glen, L.M. Leemis (eds.), *Computational Probability Applications*, International Series in Operations Research & Management Science 247,
DOI 10.1007/978-3-319-43317-2_8

8.1 Introduction

The Wilcoxon signed-rank test is an invaluable nonparametric statistical hypothesis test. Wilcoxon [166] introduced both this test and the rank-sum test in a discussion of methods of comparing treatments. The signed-rank test, used for paired data, tests the null hypothesis that the population median of the differences between pairs is equal to 0 against the alternative hypothesis that the population median of the differences is not equal to 0. The details of the test will be explained later in this paper.

The Wilcoxon signed-rank test has been studied extensively in the literature. A basic introduction to the test is in Woolson [173]. There are a number of ways to obtain tail probabilities to draw conclusions using this test. Wilcoxon [166] provides tables of significance levels for sample sizes from 7 to 16. McCornack [112] provides a recursive algorithm for generating the coefficients of the probability mass function of the test statistic under H_0. Wilcoxon [167] notes that the distribution of the test statistic under H_0 approaches a normal distribution in the limit as n goes to infinity. Bellera et al. [11] provide a number of graphics demonstrating the convergence of the distribution of the Wilcoxon statistic to a normal distribution. A good introduction to nonparametric statistics can be found in Hollander and Wolfe [68]. A discussion of non parametric methods appears in Siegel [146]. Hájek et al. [64] provide an introduction to tests involving ranks. Conover and Iman [33] discuss ranking procedures and their relationship to both parametric and non parametric statistics.

There has been very little work done in computing the exact power of the Wilcoxon signed-rank test. Büning and Qari [22] use Monte Carlo simulation to investigate the power of both the Wilcoxon test and the sign test under different configurations. Further, Monte Carlo studies include Blair and Higgins [15] and Bridge and Sawilowsky [21]. Pagano [127] plots a power curve for the sign test for slightly altered null and alternative hypotheses. Arnold [4] considers the problem of calculating power using multiple integrals. Klotz [80] uses a recursive strategy for power computations.

The goal of this paper is to analyze the distribution of the Wilcoxon test statistic and investigate the power of the Wilcoxon test. We use APPL to plot power curves and compare the power of two non parametric statistical tests—the Wilcoxon signed-rank test and the sign test. We also compare the power of the Wilcoxon test for several population distributions. Some interesting conclusions emerge.

Plotting these power curves is nontrivial. While the distribution of the test statistic under the null hypothesis is independent of the distribution of the data, the same is not true under an alternative hypothesis. The distribution of the test statistic under the null hypothesis is based on the assumption that the sign of a difference between pairs is independent of the rank of the absolute value of the difference. This assumption fails under an alternative hypothesis because the distribution is not symmetric around zero. Thus, power curves for the Wilcoxon signed-rank test can only be plotted when the distribution of the data is assumed. Tail probabilities of the distribution of the test statistic under an alternative hypothesis can be found using the joint PDF of the sample.

An algorithm is presented here that uses a computer algebra system to compute the necessary probabilities. This allows us to plot power curves for the Wilcoxon signed-rank test, compare these to power curves of the sign test, and compare how the power of the signed-rank test changes when the distribution of the data changes.

Section 8.2 contains a discussion of the Wilcoxon signed-rank test and the sign test. Section 8.3 introduces the distribution of the test statistic under H_0. Sections 8.4 and 8.5 discuss an algorithm for computing power functions for the Wilcoxon test. Section 8.6 contains a comparison of the Wilcoxon signed-rank test and the sign test and a discussion of how power is affected by the sample size and the distribution of the data. Sections 8.7–8.9 introduce and analyze the Wilcoxon–Mann–Whitney test using the symbolic capabilities of Maple and APPL. Section 8.10 contains conclusions.

8.2 Explanation of the Tests

The Wilcoxon signed-rank test is a non parametric statistical hypothesis test. It is typically used with a set of paired data in which the differences between pairs is of interest. It can also be used for the one-sample location problem under the assumption of symmetry. Let the random variables X and Y represent two observations from two continuous populations, and let the random variable $Z = X - Y$ be their difference. Let (x_i, y_i) be observed pairs of data, for $i = 1, 2, \ldots, n$. Let $z_i = x_i - y_i$ for $i = 1, 2, \ldots, n$. Let θ be the population median of Z and consider the hypothesis test

$$H_0 : \theta = 0 \qquad\qquad H_1 : \theta \neq 0.$$

One ranks the z_i values in order of increasing absolute value and then computes W^+ and W^- as the sum of the ranks of the positive and negative z_i, respectively. When ties occur, the average rank is given to all values that are tied. Consider the distribution of W^+ and W^-. Under H_0, W^+ and W^- are distributed identically. However, when one is known, the other is uniquely determined. This is because

$$W^+ + W^- = \frac{n(n+1)}{2}$$

always holds. In other words, W^+ and W^- have correlation -1. Thus, it suffices to look at only one of W^+ and W^-. Wilcoxon [166] recommends using whichever is smaller. Others recommend using W^+ or $W^+ - W^-$ as the test statistic. For this analysis, let the test statistic be $W = W^+$. Once the test statistic is observed, the null hypothesis is rejected if the probability of obtaining the observed test statistic or one more extreme under H_0 (i.e., the p-value) is smaller than a prescribed significance level α.

The distribution of Z is of interest. The only assumptions on Z are that the distribution is continuous and symmetric around a median θ. Furthermore, we assume that the sample values $Z_1, Z_2, \ldots, Z_n$ are mutually independent. In practice, one must deal with ties in the ranking of the data. However, because of the continuity assumption, this is not a concern when working with Z.

The sign test is a simpler version of the Wilcoxon signed-rank test. As before, let (x_i, y_i) be pairs of data, for $i = 1, 2, \ldots, n$, and let $z_i = x_i - y_i$ for $i = 1, 2, \ldots, n$. Let θ be the population median of the differences, and consider testing the same hypotheses as above. Instead of ranking, we simply count R^+ and R^- as the number of the observed z_i that are above and below 0, respectively. The test statistic is $R = R^+$. Again, the null hypothesis will be rejected if the probability under H_0 of obtaining the observed test statistic or one more extreme is smaller than a prescribed significance level.

For the rest of this paper, let X and Y be continuous random variables with unknown distributions unless a distribution is specified. Let the random variable $Z = X - Y$, and let θ and μ be population median and mean of Z, respectively.

8.3 Distribution of the Test Statistic Under H_0

The exact distribution of W under H_0 is mathematically intractable. However, W can be expressed as a function of n Bernoulli random variables. Under H_0, we assume that $P(Z < 0) = 1/2$ because the population median is $\theta = 0$. We assume that the probability that Z_i falls below 0 is independent of the rank of Z_i because the distribution is symmetric around the median. This is shown in Hogg et al. [67, p. 532]. Thus, the contribution to W associated with the first rank is equally likely to be 0 or 1. It is a Bernoulli(1/2) random variable. The contribution to W associated with the second rank is equally likely to be 0 or 2. It is two times a Bernoulli(1/2) random variable. Iterating this process yields the expression

$$W = \sum_{i=1}^{n} iV_i,$$

where the V_i are mutually independent and identically distributed Bernoulli(1/2) random variables. A normal approximation is often used when n is sufficiently large. The standard advice is to use the normal approximation when $n \geq 10$, e.g., Hogg et al. [67, p. 534]. The mean and variance of W under H_0 are

$$\begin{aligned} E[W] &= E[V_1 + 2V_2 + \cdots + nV_n] \\ &= (1 + 2 + \cdots + n)E[V_1] \\ &= \frac{n(n+1)}{4} \end{aligned}$$

and

$$\begin{aligned} V[W] &= V[V_1 + 2V_2 + \cdots + nV_n] \\ &= \left(1 + 2^2 + \cdots + n^2\right) V[V_1] \\ &= \frac{n(n+1)(2n+1)}{24}. \end{aligned}$$

The Wilcoxon test statistic under H_0 shows a rapid convergence to the normal distribution. The exact distribution of W can be determined using the APPL code generates the p-value of the Wilcoxon test for $n = 10$ and arbitrarily chosen test statistic $w = 46$.

```
> n := 10;
> V := BernoulliRV(1 / 2);
> for i from 2 to n do
> T := [[1 / 2, 1 / 2], [0, i], ["Discrete", "PDF"]];
> V := Convolution(V, T);
> od:
> 2 * SF(V, 46);
```

This yields a p-value $p = 33/512 \cong 0.064$. Note that the code doubles the value of the survivor function because the alternative hypothesis is two-sided. APPL can also be used to plot the exact probability mass function of the Wilcoxon statistic under H_0. This is plotted in Figure 8.1 for $n = 10$. The support of W ranges from 0 to $n(n+1)/2 = 10 \cdot 11/2 = 55$. Note that the lowest and highest three values of the probability mass function are all $1/1024$. The normal approximation, thus, seems dubious even for $n = 10$. This underscores the need for software to perform these calculations for small and moderate values of n. In the next three sections we turn our attention to using a computer algebra system to plot power curves for the Wilcoxon signed-rank test and the sign test.

8.4 Wilcoxon Power Curves for $n = 2$

We now turn to an analytical approach to plotting power curves for the Wilcoxon signed-rank test when $n = 2$ with the hopes that the same strategy can be generalized to larger sample sizes. Let $n = 2$. The sample of interest is Z_1, Z_2. We assume that Z_1 and Z_2 are independent and come from the same continuous, symmetric probability distribution. If we assume a distribution for Z, we do not need to know the distribution of X and Y to conduct the test. Because X_i and Y_i are paired, they are likely not independent. However, it makes sense to choose distributions for Z that seem likely to result from common distributions for X and Y.

As an example, assume that Z has the triangular distribution, and is symmetric around median θ. This situation seems reasonable because

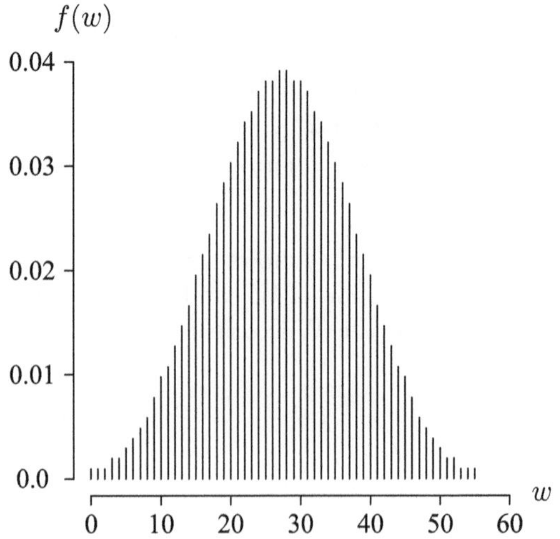

Fig. 8.1. Probability mass function when $n = 10$

the difference of two independent uniform random variables with the same width will always have a symmetric triangular distribution. Let $Z \sim \text{triangular}(-4-\delta, -\delta, 4-\delta)$ for $-4 < \delta < 4$. Under the null hypothesis, X and Y have the same median, which corresponds to $\delta = 0$. The distribution of Z satisfies the symmetry assumption with median $\theta = -\delta$. The PDF of Z is

$$f_Z(z) = \begin{cases} (z+\delta+4)/16 & -4-\delta < z < -\delta \\ (-z-\delta+4)/16 & -\delta \le z < 4-\delta. \end{cases}$$

Because Z_1 and Z_2 are independent, the joint PDF of Z_1 and Z_2 is simply the product of the marginal PDF $f_{Z_1}(z_1)$ and $f_{Z_2}(z_2)$. The joint PDF is defined on the square where both Z_1 and Z_2 range from $-4-\delta$ to $4-\delta$. It is defined in a piecewise fashion, with changes in the definition occurring at the lines $z_1 = -\delta$ and $z_2 = -\delta$. Probabilities related to these two random variables can be found by computing double integrals of the joint PDF. The support of the joint PDF of Z_1 and Z_2 is plotted in Figure 8.2 for $\delta = -2$. The support of Z_1 and Z_2 is the entire square $[-2, 6] \times [-2, 6]$.

The joint PDF of Z_1 and Z_2 for $\delta = -2$ is

$$f_{Z_1,Z_2}(z_1, z_2) = \begin{cases} (z_1z_2 + 2z_1 + 2z_2 + 4)/256 & -2 < z_1 < 2,\ -2 < z_2 < 2 \\ (-z_1z_2 + 6z_1 - 2z_2 + 12)/256 & -2 < z_1 < 2,\ 2 \le z_2 < 6 \\ (-z_1z_2 - 2z_1 + 6z_2 + 12)/256 & 2 \le z_1 < 6,\ -2 < z_2 < 2 \\ (z_1z_2 - 6z_1 - 6z_2 + 36)/256 & 2 \le z_1 < 6,\ 2 \le z_2 < 6. \end{cases}$$

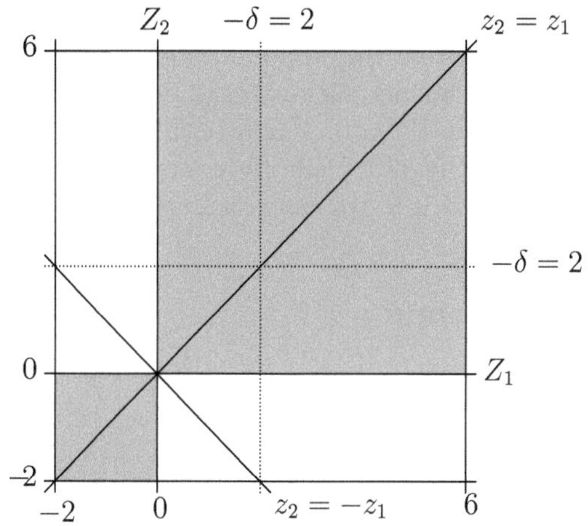

Fig. 8.2. The support of Z_1 and Z_2 for $\delta = -2$

As δ varies between -4 and 4, the square defining its support will move along the 45° line $z_2 = z_1$.

The support of the test statistic W is $\{0, 1, 2, 3\}$. Define the rejection region as $\{w \mid w \leq 0 \text{ or } w \geq 3\}$. Only two values of W will result in rejecting the null hypothesis: $W = 0$ and $W = 3$. The areas corresponding to $W = 0$ and $W = 3$ are shaded in Figure 8.2. The square to the southwest of the origin corresponds to $W = 0$; the square to the northeast of the origin corresponds to $W = 3$. There are four possible arrangements of the z_i values above and below zero. The possible arrangements are plotted in Figure 8.3.

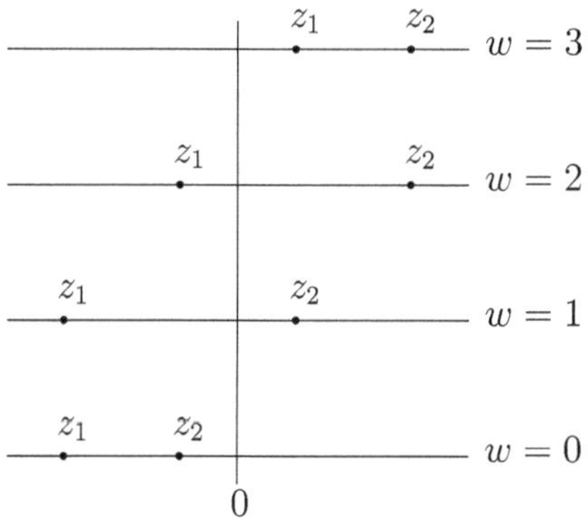

Fig. 8.3. Possible arrangements of z_1 and z_2

Ignoring the possible cases resulting from switching z_1 and z_2, two of the four cases plotted result in rejecting the null hypothesis. The probabilities of these two configurations can be calculated as double integrals. When both sample points are negative, $W = 0$. When both sample points are positive, $W = 3$. If $f_{Z_1,Z_2}(z_1, z_2)$ is the joint PDF under the alternative hypothesis that the population median of Z is θ, we can calculate the power of the test as

$$\begin{aligned} Q(\theta) &= P(\text{reject } H_0 \,|\, \theta) \\ &= P(W = 0 \,|\, \theta) + P(W = 3 \,|\, \theta) \\ &= \int_{-\infty}^{0} \int_{-\infty}^{0} f_{Z_1,Z_2}(z_1, z_2) dz_1 dz_2 + \int_{0}^{\infty} \int_{0}^{\infty} f_{Z_1,Z_2}(z_1, z_2) dz_1 dz_2. \end{aligned}$$

The integration required is complicated because the joint PDF is a piecewise function. Computing the above integrals for values of δ such that $-4 < \delta < 4$ yields the power curve plotted in Figure 8.4. The Maple code to compute the double integrals of the triangular PDF for various values of δ is given below.

```
> # Computes power over a range of delta values
> powSeq := NULL;
> for delta from -4 to 0 by 0.1 do
>    f1 := x -> 2 * (x + 4 + delta) / (8 * 4);
>    f2 := x -> 2 * (4 - delta - x) / (8 * 4);
>    W0 := 2 * int(int(f1(z1) * f1(z2),
>       z2 = -4 - delta .. z1), z1 = -4 - delta .. 0);
>    W3 := 1 / 4;
>    W3 := W3 + 2 * int(int(f1(z1) * f1(z2),
>       z2 = 0 .. z1), z1 = 0 .. -delta);
>    W3 := W3 + 2 * int(int(f2(z1) * f1(z2),
>       z2 = 0 .. -delta), z1 = -delta .. 4 - delta);
>    pow := W0 + W3;
>    powSeq := powSeq, pow;
> od:
> print(powSeq);
```

8.5 Generalization to Larger Sample Sizes

We present an algorithm for computing the power of the Wilcoxon signed-rank test against any alternative hypothesis for any sample size. Let $Z_1, Z_2, \ldots Z_n$ be the sample differences, and let $f_{Z_1, Z_2, \ldots, Z_n}(z_1, z_2, \ldots, z_n)$ be the joint PDF of the sample under a given alternative hypothesis. An algorithm for computing the power of the test against the simple alternative hypothesis $H_a : \theta = \theta_a$ is as follows.

- Enumerate all possible arrangements of the ranks above and below 0.
- Determine the corresponding value of w for each arrangement.

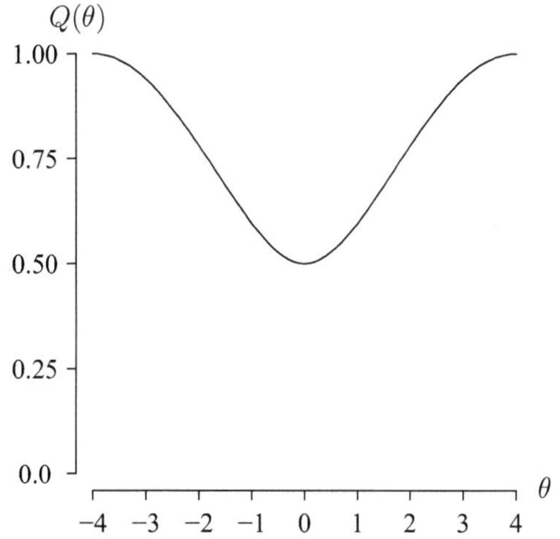

Fig. 8.4. Power function for triangular Z and $n = 2$

- Determine which arrangements correspond to a w in the rejection region.
- Arbitrarily order the z_i and integrate over proper regions of the joint PDF.
- Multiply some of the integrals by $n!$ as needed to account for switching indices, and sum the integrals.

As an example, let $n = 3$. The support of W is the set $\{0, 1, \ldots, 6\}$. Define the rejection region as $\{w \mid w \le 1 \text{ or } w \ge 5\}$. This corresponds to a significance level $\alpha = 1/2$. The power of the test is

$$\begin{aligned}
Q(\theta) &= P(\text{reject } H_0 \mid \theta) \\
&= P(W = 0 \mid \theta) + P(W = 1 \mid \theta) + P(W = 5 \mid \theta) + P(W = 6 \mid \theta) \\
&= \int_{-\infty}^{0} \int_{-\infty}^{0} \int_{-\infty}^{0} f_{Z_1,Z_2,Z_3}(z_1, z_2, z_3) dz_1 dz_2 dz_3 + \\
&\quad 6 \int_{-\infty}^{0} \int_{z_3}^{0} \int_{0}^{-z_2} f_{Z_1,Z_2,Z_3}(z_1, z_2, z_3) dz_1 dz_2 dz_3 + \\
&\quad 6 \int_{0}^{\infty} \int_{0}^{z_3} \int_{-z_2}^{0} f_{Z_1,Z_2,Z_3}(z_1, z_2, z_3) dz_1 dz_2 dz_3 + \\
&\quad \int_{0}^{\infty} \int_{0}^{\infty} \int_{0}^{\infty} f_{Z_1,Z_2,Z_3}(z_1, z_2, z_3) dz_1 dz_2 dz_3,
\end{aligned}$$

where $n! = 6$ accounts for switching the indices. Note that the integrals corresponding to $P(W = 0)$ and $P(W = 6)$ do not need to be multiplied by $n!$. In general, all the n-fold integrals except those at extreme values of W must be multiplied by $n!$. Using this algorithm and a computer algebra system, one can compute power functions for the Wilcoxon signed-rank test for any n.

The computations become intractable for large values for n because of CPU time. Assuming a uniform distribution for Z allows for extending the analysis to larger sample sizes. The PDF of the uniform distribution is constant, which makes it simple to integrate. Letting Z have a uniform distribution with width

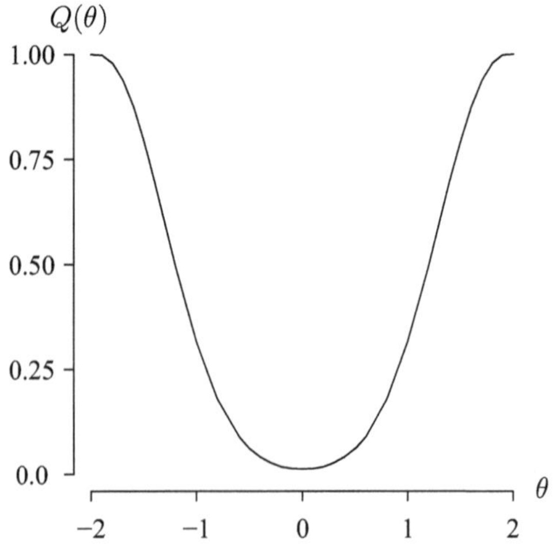

Fig. 8.5. Power function for uniform Z and $n = 10$

4 and median $-2 < \theta < 2$ yields the power curve in Figure 8.5 for $n = 10$. The rejection region is $\{w \,|\, w \le 4 \text{ or } w \ge 51\}$. This corresponds to a significance level $\alpha = 7/512 \cong 0.01367$.

In the next section, we consider a comparison of the power of the Wilcoxon signed-rank test for different sample sizes and population distributions.

8.6 Comparisons and Analysis

One would expect that, as the sample size increases, the power of the test will increase. However, it may not be obvious how much the power will increase for a given increase in the sample size. Figure 8.6 plots the power of the Wilcoxon signed-rank test when Z is normally distributed for $n = 2$ and $n = 3$ with rejection for the extreme values of W using the algorithm.

As expected, the power of the test is higher for $n = 3$ than for $n = 2$. Even for a small increase in the sample size, there is a notable increase in power. We now look at the effect the distribution of Z has on power.

Because the Wilcoxon signed-rank test is a non parametric test, the distribution of W under H_0 does not depend on the distribution of Z. Under H_1, however, the distribution of W changes depending on the assumed distribution for Z. This is because the distribution of Z is not symmetric around 0 under H_1. Thus, the assumption that the sign of Z_i is independent of the rank of Z_i does not hold. Power curves will differ for different distributions for Z.

We consider a comparison of the power of the test with two assumed distributions for Z: normal and triangular. We consider the symmetric triangular$(-4-\delta, -\delta, 4-\delta)$ distribution as before. It can be shown that the

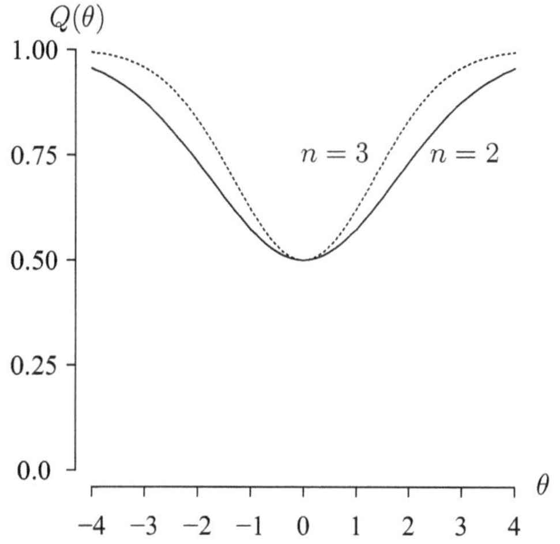

Fig. 8.6. Power functions for normal Z and $n = 2$ and $n = 3$

variance of this distribution is $V[Z] = 8/3$. To compare the two distributions on equal footing, we will consider a normal distribution with the same variance. The power curves for these two population distributions and $n = 4$ are plotted in Figure 8.7. The power curves for the two distributions are quite close. The power of the test for triangular Z is slightly higher for values of θ far from zero. The power of the test for normal Z is slightly higher for values of θ close to zero. Because the power curves are so close and the variances of the population distributions are equal, it seems that the power of the test depends more on the variance of the population than the population distribution itself. For values of θ far from zero, it is possible that the power is higher for a distribution with less weight in the tails. If the distribution of Z has less weight in the tails, it is less likely that a small number of high ranked sample points on the opposite side of 0 from the true median will cause the test statistic to fall out of the rejection region. For the case in which Z is triangular, the support of the joint distribution of the sample is confined to the square from Figure 8.2. Thus, when $|\theta| \geq 4$, the power of the test will be 1. When Z is normally distributed, the power will never reach 1 because the support is unbounded.

Next we consider a comparison of the Wilcoxon signed-rank test to the sign test discussed in Section 8.2. One would expect the Wilcoxon signed-rank test to be more powerful because it utilizes both the sign and the magnitude of the z_i values.

Assuming that Z has the uniform distribution, we can find power curves for $n = 6$ such that the significance level for both tests is $\alpha = 7/32 = 0.21875$. This is the smallest sample size for which nontrivial rejection regions can be defined for both tests with the same significance level α. The power curves for both tests are plotted together in Figure 8.8 for sampling from normal populations. As expected, the Wilcoxon signed-rank test shows higher power than the sign test.

8.7 The Wilcoxon–Mann–Whitney Test

The Wilcoxon–Mann–Whitney test goes by many names, including the Wilcoxon rank sum test and the Mann–Whitney U-test. The first description of what became the Wilcoxon–Mann–Whitney test is in Wilcoxon [166], where he describes both unpaired and paired examples of rank tests. He explains one method of calculating the exact distribution of the test statistic under the null hypothesis. Mann and Whitney [105] take a different approach with this test and transform it into the equivalent U-test. They defined a recurrence relation used to calculate the null distribution of U in the absence of ties.

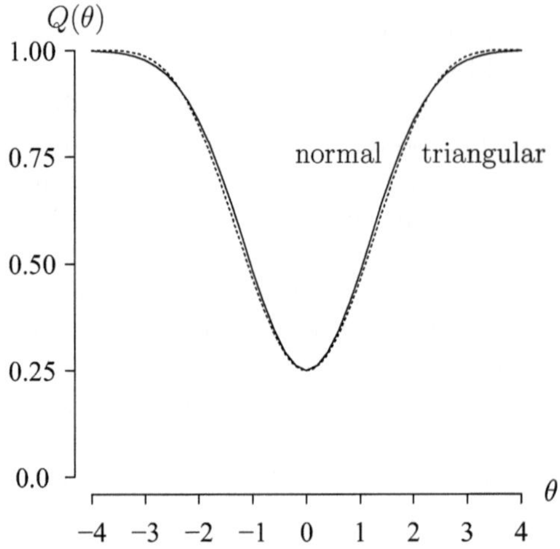

Fig. 8.7. Power functions for normal Z and triangular Z, equal variances

It is of interest to know how to compute the exact distribution of the test statistic in the Wilcoxon–Mann–Whitney test. Mundry and Fischer [116] explain that statistical programs at the time would use asymptotic tests for small samples when an exact procedure was more appropriate. One problem that arises in this test is the presence of ties in the samples. Bergman et al. [12] found significant differences across various statistical packages in regards to how the Wilcoxon–Mann–Whitney test is dealt with. They conducted the test on a data set containing ties on 11 different statistical programs and found that the data was handled very differently in each one. Of the 11 programs,

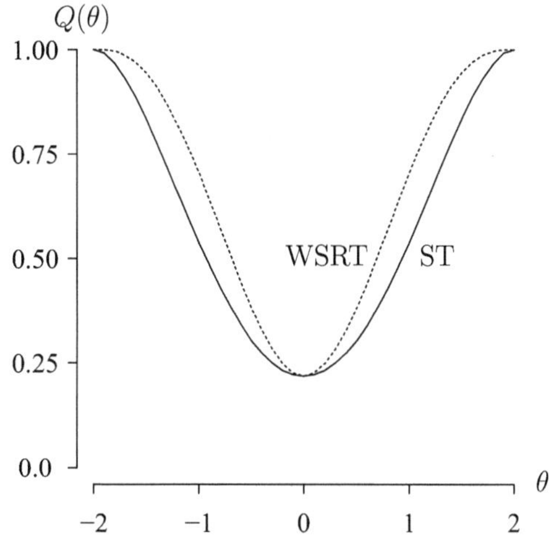

Fig. 8.8. Power curves for the sign test (ST) and the Wilcoxon signed-rank test (WSRT) for $n = 6$

only four used an exact permutation. Most programs corrected for ties in some manner and others performed pseudo-exact calculations. It is unclear what the exact distribution of the Wilcoxon–Mann–Whitney test statistic is in the presence of ties.

8.8 Explanation of the Test

Similar to the Wilcoxon signed-rank test, the Wilcoxon–Mann–Whitney rank sum test is a non parametric statistical hypothesis test. It is used with two random samples of data that are not paired; therefore, the two samples do not need to be of equal sizes. The test determines whether the samples could have come from the same population. Let $x_1, x_2, \ldots, x_m$ be the data values associated with a random sample drawn from the first population; let $y_1, y_2, \ldots, y_n$ be the data values associated with a random sample from the second population. We assume without loss of generality that $m \leq n$. Let F_X denote the CDF of the population; let F_Y denote the CDF of the second population. A two-sided hypothesis test is most commonly performed. The associated null and alternative hypotheses are

$$H_0 : F_X = F_Y \qquad\qquad H_1 : F_X \neq F_Y.$$

The samples are combined and ranked in increasing value from 1 to N, where $N = m + n$. If ties are present, the average rank is given to the tied values. Because this test goes by many names, there are many variants of the test statistic. Let R_{x_i} be the ranks of the data values from the first population.

Due to its simplicity, the following test statistic will be used: $W = \sum_{i=1}^{m} R_{x_i}$, i.e., W is the sum of the ranks of the data values from the first population.

Most texts refer to tables for determining the rejection region for W. These tables are only accurate for cases when no ties are present. When ties are present, statistical packages often use a normal or chi square approximation for the test to determine the p-value, regardless of sample size. This method is also implemented for large sample sizes (though what constitutes a large sample size varies depending on the software).

8.9 Three Cases of the Distribution of W Under H_0

The exact distribution for W under H_0 is of interest. The distribution without ties is quite simple to obtain (even for large sample sizes given a computer with enough power). However, the distribution of W in the presence of ties is much more difficult to determine. Using simple examples with samples sizes $m = 2$ and $n = 3$, the following cases will be addressed: no ties, ties present within one sample exclusively, and ties present between both samples.

8.9.1 Case I: No Ties

To determine the distribution of W without ties, we simply have to calculate the frequencies of the sums of all combinations of the ranks. Consider the following two samples with $m = 2$ and $n = 3$:

Sample from population 1	7.5 5.6
Sample from population 2	3.0 4.9 6.6

Since the smaller sample is from the first population, W is sum of the ranks of x_1 and x_2. The first step to finding the distribution of W is to combine the samples and rank them.

Sample	y_1	y_2	x_2	y_3	x_1
Value	3.0	4.9	5.6	6.6	7.5
Rank	1	2	3	4	5

We can see that the ranks for x_1 and x_2 are 5 and 3, so $W = 8$. Now let's look at the probability of obtaining $W = 8$.

There are $\binom{N}{m} = \binom{5}{2} = 10$ possible ways to select two ranks for the two values from the first population from the rank values 1 to 5. Since N and m are small, we can enumerate all ten possible pairs and their sums:

Combinations	(1, 2)	(1, 3)	(1, 4)	(1, 5)	(2, 3)	(2, 4)	(2, 5)	(3, 4)	(3, 5)	(4, 5)
Sums	3	4	5	6	5	6	7	7	8	9

Under H_0, each of these pairs of ranks is equally likely, so we can tabulate the probability mass function of W under H_0.

Sums	3	4	5	6	7	8	9
Frequency	1	1	2	2	2	1	1
Probability	0.1	0.1	0.2	0.2	0.2	0.1	0.1

This probability mass function is plotted in Figure 8.9. The probability mass function of W is symmetric for all values of m and n, with minimum support value $m(m+1)/2$ and maximum support value $N(N+1)/2 - n(n+1)/2$.

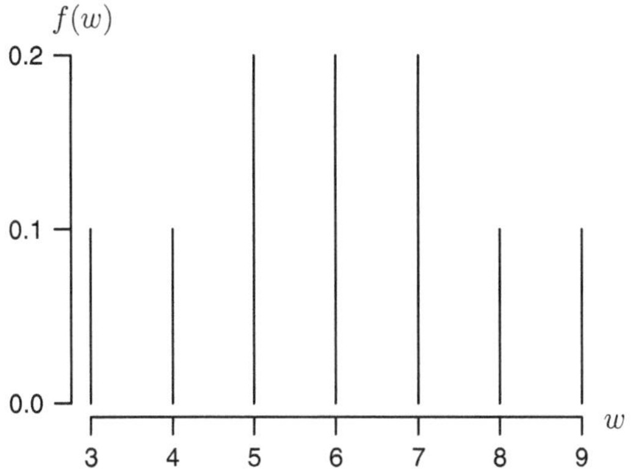

Fig. 8.9. Probability mass function when $m = 2$ and $n = 3$ with no ties in the samples

We now have the distribution of W under H_0. We can calculate a p-value for a two-sided hypothesis test with $\alpha = 0.20$. The test is two sided and, given that our distribution is symmetric, our p-value is $\mathrm{P}(W \geq 8) + \mathrm{P}(W \leq 4) = 0.40$. Since our p-value is large, we fail to reject the null hypothesis.

We can calculate this p-value in R using the `exactRankTests` library with the code below:

```
> library(exactRankTests)
> x = c(7.5, 5.6)
> y = c(3.0, 4.9, 6.6)
> wilcox.exact(x, y, paired = FALSE, alternative = "two.sided",
               exact = TRUE, conf.level = 0.80)
# Exact Wilcoxon rank sum test
# data:  x and y
# W = 5, p-value = 0.4
# alternative hypothesis: true mu is not equal to 0
```

This yields a p-value of 0.40 as expected and calculates the test statistic as $W = 5$. The difference in the two test statistics is due to a variant of the test statistic in which the smallest possible value for W is subtracted from $\sum R_{x_i}$. This results in the distribution of W having support starting at 0 rather than $m(m+1)/2$. In this example, the test statistic in R is $\sum R_{x_i} - m(m+1)/2 = 8 - 3 = 5$. The two test statistics are equivalent.

Using Table 21 from Kanji [76], which uses our form of the test statistic, the critical value for the two-sided test at our given α is 3. So test statistic values less than or equal to 3 and greater than or equal to 9 would result in a rejection of the null hypothesis. However, our test statistic $W = 8$ is in between the two values, so we again fail to reject the null hypothesis.

The general algorithm for obtaining the distribution of W under H_0 without ties is to find all $\binom{N}{m}$ combinations, sum the combination values, and divide the frequencies of these sums by the number of total combinations $\binom{N}{m}$. The following APPL code calculates this given values for m and n and returns in the correct list-of-lists format for random variables.

```
> WMWRV := proc(m :: posint, n :: posint)
> local N, mmax, nmax, Nmax, bino, Supp, Prob, Comb, i, j,
>   sumc, ListOfLists:
> N := m + n:
> mmax := m * (m + 1) / 2:
> nmax := n * (n + 1) / 2:
> Nmax := N * (N + 1) / 2:
> bino := binomial(N, m):
> Supp := []:
> Prob := []:
> for i from mmax to (Nmax - nmax) do
>   Supp := [op(Supp), i]:
>   Prob := [op(Prob), 0]:
> od:
> Comb := []:
> for i from 1 to m do
>   Comb := [op(Comb), i]:
> od:
> for i from 1 to bino do
>   sumc := add(Comb[j], j = 1 .. m):
>   Prob[sumc - mmax + 1] := Prob[sumc - mmax + 1] + 1 / bino:
>   Comb := NextCombination(Comb, N):
> od:
> ListOfLists := [Prob, Supp, ["Discrete", "PDF"]]:
> RETURN(ListOfLists):
> end:
> WMWRV(2, 3);
```

8.9.2 Case II: Ties Only Within Each Sample

Though ties do present a problem for calculating the distribution W, when the ties are present within individual samples, they are easier to deal with. Consider the following two samples with $m = 2$ and $n = 3$:

Sample from population 1	5.6 5.6
Sample from population 2	3.0 4.9 6.6

In theory, tied observations will not occur for samples drawn from continuous distributions, but due to limitations of the precision of measurements and rounding, ties occur quite often. Therefore it would not be unusual to see a tie such as the one in the first sample. Continuing with the example, rank the values next.

Sample	y_1	y_2	x_1	x_2	y_3
Value	3.0	4.9	5.6	5.6	6.6
Rank	1	2	3.5	3.5	5

The test statistic for this data set is $w = 7$. If we had not taken the average ranks of the tied values and assuming two integer-valued ranks are assigned at random, the test statistic would still be $w = 7$ because the ranks for x_1 and x_2 would have been 3 and 4. From here, there are two ways to assess the distribution. First, because the test statistic is the same value it would have been without accounting for ties, the distribution of W can be viewed as the same as it was in our previous example in Figure 8.9. We have a p-value of $\mathrm{P}(W \leq 5) + \mathrm{P}(W \geq 7) = 0.80$. The second way to look at this distribution is to make a conditional distribution dependent on the data. As before, we have ten combinations and their sums.

Combinations	(1, 2)	(1, 3.5)	(1, 3.5)	(1, 5)	(2, 3.5)	(2, 3.5)	(2, 5)	(3.5, 3.5)
Sums	3	4.5	4.5	6	5.5	5.5	7	7
Combinations	(3.5, 5)	(3.5, 5)						
Sums	8.5	8.5						

We tabulate the frequencies and plot the corresponding probability mass function in Figure 8.10.

Sums	3	4.5	5.5	6	7	8.5
Frequency	1	2	2	1	2	2
Probability	0.1	0.2	0.2	0.1	0.2	0.2

Using the R code below, we get the p-value:

```
> x = c(5.6, 5.6)
> y = c(3.0, 4.9, 6.6)
> wilcox.exact(x, y, paired = FALSE,
>          alternative = "two.sided", exact = TRUE)
# Exact Wilcoxon rank sum test
# data:  x and y
# W = 4, p-value = 0.7
# alternative hypothesis: true mu is not equal to 0
```

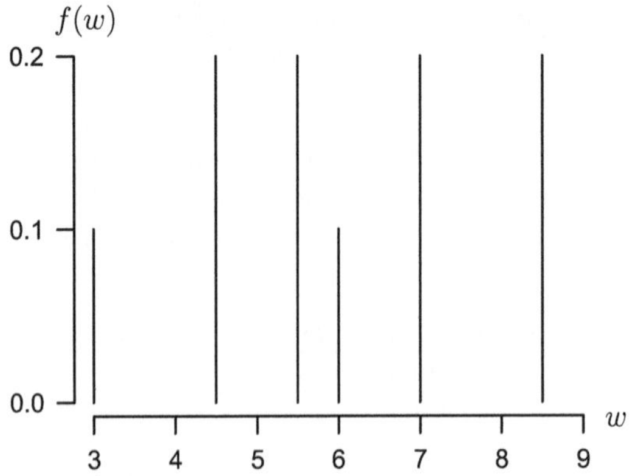

Fig. 8.10. Probability mass function when $m = 2$ and $n = 3$ with ties within a sample

Though the conditional distribution of W is not symmetric, R still calculates the p-value in the same way. So in this case, $P(W \leq 5) + P(W \geq 7) = 0.70$. Again, the test statistic computed by R is just a shifted version of the test statistic defined here, so the two are equivalent. The `wilcox.exact` function uses the shift-algorithm formulated by Streitberg and Rohmel [155], which calculates the same conditional distribution that is presented here. In both scenarios for the distribution of W with a tie within one sample, we fail to reject the null hypothesis. Using the conditional distribution yielded a smaller p-value, which may have resulted in a different conclusion in a hypothesis test had it been small enough.

8.9.3 Case III: Ties Between Both Samples

Now that we have seen the simplest example of a tie within a single sample, consider the case in which there is a tied value that is common between the two samples.

Sample from population 1	7.5	5.6	
Sample from population 2	5.6	4.9	6.6

The tied value, 5.6, is present in both samples. We can calculate the conditional distribution given the data values as we did in our previous example, but our test statistic will no longer necessarily be an integer.

Sample	y_2	x_2	y_1	y_3	x_1
Value	4.9	5.6	5.6	6.6	7.5
Rank	1	2.5	2.5	4	5

Since we have two values of 5.6, one in each sample, we cannot say that the test statistic will remain the same regardless of ties because our tied value from population 1 could have a rank of 2 or 3 before it was averaged. The test statistic is $W = 7.5$. As in the previous example, we can calculate the conditional probability distribution based on the data.

Combinations	(1, 2.5)	(1, 2.5)	(1, 4)	(1, 5)	(2.5, 2.5)	(2.5, 4)	(2.5, 5)
Sums	3.5	3.5	5	6	5	6.5	7.5
Combinations	(2.5, 4)	(2.5, 5)	(4, 5)				
Sums	6.5	7.5	9				

We can now determine the conditional distribution in the table below. The probability mass function is plotted in Figure 8.11.

Sums	3.5	5	6	6.5	7.5	9
Frequency	2	2	1	2	2	1
Probability	0.2	0.2	0.1	0.2	0.2	0.1

The R code below calculates the p-value.

```
> x = c(7.5, 5.6)
> y = c(5.6, 4.9, 6.6)
> wilcox.exact(x, y, paired = FALSE, alternative = "two.sided",
>         exact = TRUE)
# Exact Wilcoxon rank sum test
# data:  x and y
# W = 4.5, p-value = 0.5
# alternative hypothesis: true mu is not equal to 0
```

As with the previous conditional distribution, though it is not symmetric, the p-value is calculated the same way: $\mathrm{P}(W \geq 7.5) + \mathrm{P}(W \leq 4.5) = 0.50$.

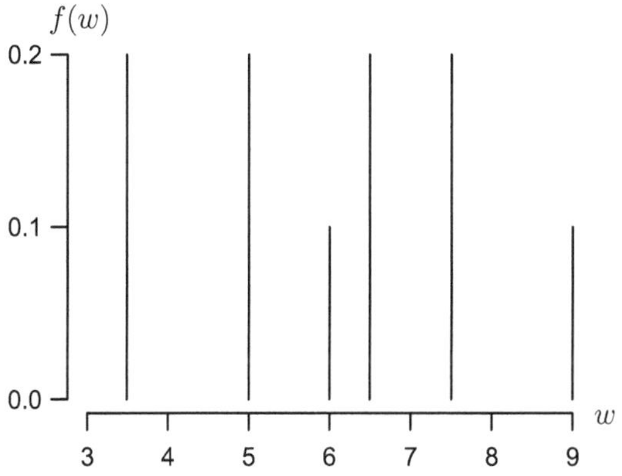

Fig. 8.11. Probability mass function when $m = 2$ and $n = 3$ with ties present between samples

8.10 Conclusions

We have presented notes on two rank statistics. We have presented an algorithm for computing the power of the Wilcoxon signed-rank test for small sample sizes. If computational power were available, this algorithm should generalize to larger sample sizes. The power of the Wilcoxon signed-rank test changes when the assumed distribution of Z changes. Furthermore, there is a relationship between the power of the test and the weight in the tails of the assumed distribution. As expected, the Wilcoxon test is more powerful than the sign test. However, the degree to which the Wilcoxon test is more powerful changes with the assumed distribution. For the Mann–Whitney test, we have determined distributions under the null hypothesis for the case of ties within and between samples. Future work in this area could be done to increase the sample size for which these methods are efficient. There are also a wide variety of statistical tests that could benefit from work with computer algebra systems. Computing the power of non parametric tests is not a trivial matter because of the distribution-free assumption. There is a great future in the use of computer algebra systems to make difficult statistical computations more tractable.

Acknowledgments

This research is partially supported by an NSF CSUMS grant DMS-0703532 and NSF grant number DUE-0849955 "SCORE: Scientific Computing and Operations Research Education," at The College of William & Mary.

9

Control Chart Constants for Non-normal Sampling

William H. Kaczynski and Lawrence M. Leemis

Abstract Statistical process control chart constants are bias correction factors used to establish three-sigma limits that are used to identify assignable variation. These constants have been tabulated for normal sampling. Subsequent research has verified the robustness for non-normal sampling. This paper explores exact results for both the normal distribution and select non-normal distributions using a computer algebra system to compute the exact values of the constants.

Keywords Normal sampling • Process control • Statistical quality control

9.1 Introduction

Control charts are widely used in industry to provide insight on process behavior and identify assignable causes associated with a shift in the mean value of the process. These charts were first proposed by Walter Shewhart in 1923 at Bell Telephone Laboratories [145]. To create the control limits, estimates

This original work highlights the wide application that APPL enjoys. The ability to find exact control chart constants for non-normal distributions highlights the utility of exploring many distributions in an applied setting. The calculation of control chart constants is critical in the area of quality control known as statistical process control. Use of APPL procedures `OrderStat`, `RangeStat` and `Transform` are critical to this research. Furthermore the `BootstrapRV` procedure extends such investigations into non parametric, bootstrap based analysis.

W.H. Kaczynski
US Army, Fort Leavenworth, KS, USA

L.M. Leemis (✉)
The College of William and Mary, Williamsburg, VA, USA
e-mail: leemis@math.wm.edu

A.G. Glen, L.M. Leemis (eds.), *Computational Probability Applications*, International Series in Operations Research & Management Science 247,
DOI 10.1007/978-3-319-43317-2_9

for the mean and standard error of the data's population are required, along with constants that serve as bias correction factors. The first control chart constants, then denoted by d_2 and d_3 (for the sample range), were proposed by Tippett [161]. McKay and Pearson [111] obtained the exact distribution of the sample range for $n = 3$ observations drawn from a normal distribution. Pearson and Hartley [130] tabulated the fractiles of the mean of the sample range for $n = 2$ to $n = 20$, see also Wheeler [165]. The terms *bias correction factor* and *control chart constant* are used interchangeably.

Bias correction factors for standard deviations followed a similar development. They too are based on an underlying normal distribution. For both sets of constants, extensive work exists, such as Wheeler [165], showing the robustness of these constants for data from non-normal distributions. For the most part, similar constants for non-normal distributions do not appear in the literature for two reasons: (1) most applications involve sampling from normal populations, and (2) they are not easily computed. The purpose of this paper is to offer an alternative method of computation using APPL to compute exact values of these control chart constants. Additionally, APPL typically provides exact results rather than approximations. Although normal sampling can be assumed in the vast majority of statistical process control applications, occasions will arise where non-normal sampling is an appropriate assumption. The development here allows an engineer to easily obtain the appropriate control chart constants in these alternate settings.

9.2 Constants d_2, d_3

The aforementioned constants d_2 and d_3 relate to the distribution of the sample range, denoted by R. The correction factor d_2 accounts for the discrepancy between the mean of the sample range and the population standard deviation. Given a random sample $X_1, X_2, \ldots, X_n$ from a population with CDF $F(x)$, PDF $f(x)$, finite unknown variance σ_X^2, and associated order statistics $X_{(1)}, X_{(2)}, \ldots, X_{(n)}$, the sample range, R, is

$$R = X_{(n)} - X_{(1)}. \tag{9.1}$$

The joint PDF of the order statistics $X_{(i)}$ and $X_{(j)}$ associated with a sample size n given by Hogg et al. [67] is

$$\begin{aligned} f_{X_{(i)},X_{(j)}}(x_{(i)}, x_{(j)}) = {} & \frac{n!}{(i-1)!(j-i-1)!(n-j)!} \left[F(x_{(j)}) - F(x_{(i)})\right]^{j-i-1} \\ & \times \left[F(x_{(i)})\right]^{i-1} \left[1 - F(x_{(j)})\right]^{n-j} \\ & \times f(x_{(i)}) f(x_{(j)}) \qquad x_{(i)} < x_{(j)} \end{aligned}$$

for integers $1 \le i < j \le n$ and for $i = 1$, $j = n$, this simplifies to

$$f_{X_{(1)},X_{(n)}}(x_{(1)}, x_{(n)}) = n(n-1)\left[F(x_{(n)}) - F(x_{(1)})\right]^{n-2} \times f(x_{(1)})f(x_{(n)}) \qquad x_{(1)} < x_{(n)}. \tag{9.2}$$

Burr [23] uses a change of variable, $X_{(n)} = X_{(1)} + R$ (since, by definition $R = X_{(n)} - X_{(1)}$) in Eq. (9.2) to find the joint PDF of $X_{(1)}$ and R and then integrates out $X_{(1)}$ to find the PDF of R. This, of course, works well for distributions with closed-form CDFs; however, CDFs involving mathematically intractable integrals are problematic. Once the distribution of R is obtained, it is used it to correct bias by

$$E[R] = d_2 \sigma_X. \tag{9.3}$$

Burr [23] also suggests an easier approach to find $E[R]$, which lends itself well to implementation in APPL. Using Eq. (9.1), for a sample of size n, the expected value of the sample range is

$$E[R] = E\left[X_{(n)}\right] - E\left[X_{(1)}\right], \tag{9.4}$$

therefore, using Eqs. (9.3) and (9.4), we can express d_2 as

$$d_2 = \frac{E[R]}{\sigma_X} = \frac{E\left[X_{(n)}\right] - E\left[X_{(1)}\right]}{\sigma_X}.$$

This result can be implemented using the APPL **RangeStat** procedure for select distributions. This procedure returns the distribution of the sample range for a sample of size n. Equivalently, we can use the **OrderStat** procedure, and return d_2 values exactly. For sampling from a normally distributed population, we can always remove the mean by subtraction, resulting in a random variable with mean zero. For $n = 3$ consider the APPL statements

```
> n := 3:
> X := NormalRV(0, sigma):
> (Mean(OrderStat(X, n, n)) - Mean(OrderStat(X, n, 1)))
    / sqrt(Variance(X));
```

which yield the exact value of $d_2 = 3/\sqrt{\pi}$. Though this is convenient, APPL is only capable of returning the exact symbolic expression of d_2 for $n = 2$ and $n = 3$. For $n > 3$, the problem is mathematically intractable and the integrals must be solved numerically. However, if population distribution parameter values are input for the code above APPL is capable of solving for d_2 when $n \ge 3$. Since d_2 depends only on n (and is independent of μ, σ), assigning values to these distribution parameters does not affect d_2.

We will proceed in a similar manner for d_3, which corrects for the standard deviation of the range. The relationship is

$$\sigma_R = d_3 \sigma_X \qquad \Rightarrow \qquad d_3 = \frac{\sigma_R}{\sigma_X}.$$

Since APPL can compute the exact distribution of R, we can also obtain σ_R easily for select distributions.

Example 9.1. Given that X_1, X_2, and X_3 are iid exponential(λ) random variables, find the bias correction factors d_2 and d_3 for the sample range. The APPL statements

```
> n  := 3:
> X  := ExponentialRV(lambda):
> R  := RangeStat(X, n):
> d2 := Mean(R) / sqrt(Variance(X));
> d3 := sqrt(Variance(R)) / sqrt(Variance(X));
```

yield

$$d_2 = 3/2 \qquad \text{and} \qquad d_3 = \sqrt{5/2} \cong 1.118.$$

Likewise, when $n = 18$,

$$d_2 = \frac{42142223}{12252240} \cong 3.440 \qquad \text{and} \qquad d_3 = \frac{\sqrt{238357395880861}}{12252240} \cong 1.260.$$

Table 9.1 compares values for d_2 and d_3, given the sample is drawn from exponential, normal, Rayleigh, and $U(0,1)$ distributions for sample sizes $n = 2$ to $n = 20$. These constants do not depend on the rate parameter λ (for the exponential and Rayleigh distributions) nor μ or σ (for the normal distribution).

As shown in Table 9.1, APPL is able to calculate exact values of d_2 and d_3 for the exponential, Rayleigh, and standard uniform distributions. All other distributions required numerical integration with estimated values of the parameters. So, in theory, we could estimate d_2 and d_3 for any arbitrary sampling distribution. While this might be novel, it really is not special to APPL because we are really using Maple's capability to estimate the result with numerical integration. If we do provide numeric values for parameters, we can take advantage of APPL to calculate the constants. In some cases, as illustrated in Example 9.2, APPL provides exact results.

There may be applications (e.g., life testing associated with bulbs or fuses) where a non-normal distribution is appropriate, and this provides an easy way to calculate control chart constants. Additionally, Tadikamalla et al. [159] substantiate non-normal applications providing examples that calculate the upper and lower control limits for the logistic and Laplace distributions. Though they only consider symmetric distributions, the same practice can be considered for non symmetric cases using APPL, with an added advantage of never referring to a chart calculated for specific values of n and kurtosis estimates.

Example 9.2. Given that X_1 and X_2 are iid Weibull(2, 3) random variables, find the bias correction factor d_2 for the sample range. The APPL statements

```
> n  := 2:
> X  := WeibullRV(2, 3):
> d2 := Mean(OrderStat(X, n, n))
>         - Mean(OrderStat(X, n, 1))) / sqrt(Variance(X)):
```

yield

$$d_2 = \frac{\pi\sqrt{6}\left[\left(4 - 2^{2/3}\right)/18 - 2^{2/3}\sqrt{3}/6\right]}{\sqrt{9\Gamma(2/3)^3 - 2\pi^2}} \cong 1.135.$$

The APPL procedure `OrderStat(X, n, r)` computes the exact distribution of the rth order statistic drawn from a sample of size n drawn from a population described by the random variable X.

In order to find d_2 and d_3 from first principles (as provided by Wheeler [165]) given an underlying parametric distribution, we must assign values to the distribution parameters. Even with small sample sizes, the process control literature provides well-established parameter estimation methods. However, given the normal distribution's wide acceptance in process control, current literature focuses on the normal distribution's mean μ and standard deviation σ, potentially suggesting an area of further work. Conceivably, if we knew enough about the observed process data to use a non-normal parametric model, we should also be confident in estimating the distribution's parameters. Thus, APPL provides an efficient foundation for calculating d_2 and d_3.

Selecting a distribution to adequately model observed data has many troubling issues. If the researcher does not want to make assumptions accompanying a certain parametric distribution nor introduce potential error in selection, he or she can also create a distribution via bootstrapping with well-established statistical properties established by Efron and Tibshirani [49]. Once a PDF is created using bootstrapping, APPL can compute the constants d_2 and d_3 as shown in Example 9.3 using the `BootstrapRV` procedure.

Example 9.3. Given the arbitrary PDF $f_X(x)$ created by bootstrapping for the observed order statistics $x_{(1)} = 1$, $x_{(2)} = 3$, $x_{(3)} = 4$, and $x_{(4)} = 7$, compute the constants d_2 and d_3 for sample size $n = 3$. The APPL statements

```
> data := [1, 3, 4, 7]:
> X    := BootstrapRV(data):
> R    := RangeStat(X, 3):
> d2   := Mean(R) / sqrt(Variance(X));
> d3   := sqrt(Variance(R)) / sqrt(Variance(X));
```

yield

$$d_2 = 19\sqrt{3}/20 \cong 1.645 \qquad \text{and} \qquad d_3 = \sqrt{2637}/60 \cong 0.856.$$

Table 9.1. Comparison of d_2 and d_3 for exponential, normal, Rayleigh, and $U(0,1)$ sampling distributions

	d_2				d_3			
n	Expon	Normal	Rayleigh	$U(0,1)$	Expon	Normal	Rayleigh	$U(0,1)$
2	1.000	1.128	1.121	1.155	1.000	0.853	0.863	0.816
3	1.500	1.693	1.681	1.732	1.118	0.888	0.897	0.775
4	1.833	2.059	2.041	2.078	1.167	0.880	0.885	0.693
5	2.083	2.326	2.300	2.309	1.193	0.864	0.866	0.617
6	2.283	2.534	2.501	2.474	1.210	0.848	0.848	0.553
7	2.450	2.704	2.663	2.598	1.221	0.833	0.830	0.500
8	2.593	2.847	2.797	2.694	1.230	0.820	0.815	0.455
9	2.718	2.970	2.912	2.771	1.235	0.808	0.802	0.418
10	2.829	3.078	3.012	2.834	1.241	0.797	0.790	0.386
11	2.929	3.173	3.100	2.887	1.245	0.787	0.779	0.358
12	3.020	3.258	3.179	2.931	1.248	0.778	0.769	0.334
13	3.103	3.336	3.250	2.969	1.251	0.770	0.759	0.313
14	3.180	3.407	3.314	3.002	1.253	0.762	0.752	0.294
15	3.252	3.472	3.373	3.031	1.255	0.755	0.745	0.278
16	3.318	3.532	3.427	3.057	1.257	0.749	0.738	0.263
17	3.381	3.588	3.477	3.079	1.259	0.743	0.731	0.250
18	3.440	3.640	3.524	3.099	1.260	0.738	0.726	0.238
19	3.495	3.689	3.568	3.118	1.261	0.733	0.720	0.227
20	3.548	3.735	3.608	3.134	1.263	0.729	0.715	0.217

9.3 Constants c_4, c_5

Similar to d_2 and d_3, the control chart constants c_4 and c_5 are also bias correction factors. However, as d_2 and d_3 corrected for the mean and standard deviation of the sample range R, c_4 and c_5 correct for the mean of the sample standard deviation, S, and its standard error. This is unusual because we usually discuss a sample's mean and standard deviation, but we are now focused on the *sample's mean standard deviation* and *the variance of the standard deviation.* We denote the mean of the standard deviation by μ_S and its standard deviation by σ_S. Thus, the relationships are

$$\mu_S = E\left[S\right] = c_4\sigma_X \tag{9.5}$$

and

$$\sigma_S = \sqrt{\text{Var}(S)} = c_5\sigma_X. \tag{9.6}$$

9.3.1 Normal Sampling

The derivations of c_4 and c_5 are based on the fact that $E\left[S^2\right] = \sigma_X^2$ and the well-known result

$$\frac{(n-1)S^2}{\sigma_X^2} \sim \chi^2_{n-1} \tag{9.7}$$

for normal sampling (Hogg et al. [67]), where χ^2_{n-1} denotes a chi square random variable with $n-1$ degrees of freedom. The mean of the sample standard deviation is

$$\begin{aligned}
c_4\sigma_X &= E\left[S\right] \\
&= E\left[\sqrt{S^2}\right] \\
&= E\left[\sqrt{S^2\frac{n-1}{n-1}\cdot\frac{\sigma_X^2}{\sigma_X^2}}\right] \\
&= E\left[\frac{\sigma_X}{\sqrt{n-1}}\sqrt{\frac{(n-1)S^2}{\sigma_X^2}}\right] \\
&= \frac{\sigma_X}{\sqrt{n-1}}\,E\left[\sqrt{\chi^2_{n-1}}\right].
\end{aligned}$$

Solving for c_4 yields

$$c_4 = \frac{E\left[\chi_{n-1}\right]}{\sqrt{n-1}},$$

where χ_{n-1} denotes a chi random variable with $n-1$ degrees of freedom. The standard deviation of the sample standard deviation is

$$\begin{aligned}
c_5\sigma_X &= \sigma_S \\
&= \sqrt{\operatorname{Var}\left[S\right]} \\
&= \sqrt{E\left[S^2\right]-\left[E\left[S\right]\right]^2} \\
&= \sqrt{\sigma_X^2 - E\left[S\right]E\left[S\right]} \\
&= \sqrt{\sigma_X^2 - \frac{\sigma_X^2}{n-1}E\left[\sqrt{\chi^2_{n-1}}\right]^2} \\
&= \sigma_X\sqrt{1-\frac{E\left[\chi_{n-1}\right]^2}{n-1}}.
\end{aligned}$$

Solving for c_5 yields

$$c_5 = \sqrt{1-\frac{E\left[\chi_{n-1}\right]^2}{n-1}}.$$

The result provided in Eq. (9.7) yields a distinct advantage for finding c_4 and c_5 in the normal sampling case. We can use APPL to perform the calculations independently of the parameters σ and μ, producing the exact results for c_4 and c_5 which depend only on the sample size n. The procedure `c4(n)`, written in APPL, is given below. A similar procedure, `c5(n)`, was written for c_5.

```
> c4 := proc(n)
>   local X, c4;
>   X := ChiRV(n - 1):
>   c4 := Mean(X) / sqrt(n - 1):
>   return(c4);
> end proc;
```

A call to `c4` and `c5` with the argument $n = 4$, for example, yields the exact values

$$c_4 = \frac{2\sqrt{6}}{3\sqrt{\pi}} \cong 0.921 \qquad \text{and} \qquad c_5 = \frac{1}{3}\sqrt{9 - \frac{24}{\pi}} \cong 0.389.$$

These symbolic expressions are somewhat novel in that these constants are typically tabulated in decimal form rather than exactly in symbolic form. Furthermore, to illustrate the value of APPL and Maple's symbolic computational ability in this context, consider the unlikely large sample size $n = 100$. A call to `c4(100)` produces

$$c_4 = \frac{39614081257132168796771975168\sqrt{22}}{105095150568296034723763017975\sqrt{\pi}} \cong 0.997.$$

The associated exact expression for c_5 is much too large to fit here, but the numerical value is $c_5 \cong 0.071$. The CPU time to compute these constants is negligible.

9.3.2 Non-normal Sampling

Given that observations $X_1, X_2, \ldots, X_n$, are sampled from a non-normal distribution calculating c_4 and c_5 is much more complicated. We first derive a general form of each, then investigate its calculation for select distributions.

Using Eq. (9.5), the general derivation of c_4 is

$$\begin{aligned} c_4\sigma_X &= E\left[S\right] \\ &= E\left[\sqrt{S^2}\right] \\ &= E\left[\sqrt{\frac{1}{n-1}\sum_{i=1}^{n}\left(X_i-\bar{X}\right)^2}\right] \\ &= \frac{1}{\sqrt{n-1}}\,E\left[\sqrt{\sum_{i=1}^{n}X_i^2-2n\bar{X}^2+n\bar{X}^2}\right] \\ &= \frac{1}{\sqrt{n-1}}\,E\left[\sqrt{\sum_{i=1}^{n}X_i^2-n\bar{X}^2}\right] \\ &= \frac{1}{\sqrt{n-1}}\,E\left[\sqrt{\sum_{i=1}^{n}X_i^2-\left[\sum_{i=1}^{n}X_i\right]^2\Big/n}\right]. \end{aligned}$$

Therefore, we calculate c_4 as

$$c_4=\sigma_X^{-1}\left[\frac{1}{\sqrt{n-1}}\,E\left[\sqrt{\sum_{i=1}^{n}X_i^2-\left[\sum_{i=1}^{n}X_i\right]^2\Big/n}\right]\right]. \tag{9.8}$$

In a similar manner, and using Eq. (9.6), it can be shown that a general expression for c_5 is

$$c_5=\sigma_X^{-1}\sqrt{\sigma_X^2-\frac{1}{n-1}\left(E\left[\sqrt{\sum_{i=1}^{n}X_i^2-\left(\sum_{i=1}^{n}X_i\right)^2\Big/n}\right]\right)^2}.$$

Burr [25] also presents c_5 in terms of c_4 via the relationship

$$c_5=\sqrt{1-c_4^2}.$$

Therefore, if we are successful in finding c_4 we can easily evaluate c_5, narrowing the focus of evaluation to c_4. Substituting $n=2$ into 9.8, we conclude that the numerator, $E[S]$, is

$$E\left[S\right]=\frac{1}{\sqrt{2}}\,E\left[\,|X_1-X_2|\,\right].$$

The bias correction factor is then calculated via

$$c_4 = E\left[S\right]/\sigma_X = \frac{1}{\sqrt{2}\sigma_X}\, E[\,|X_1 - X_2|\,].$$

Given that the parameter σ_X appears in the denominator of the expression, we require it to also appear in the numerator forcing a cancellation and a numerical c_4 value that is independent of σ_X. Unfortunately, this only occurs for distributions in which a single parameter involving the standard deviation appears. The next example highlights such an occurrence.

Example 9.4. Given that X_1, and X_2 are iid exponential(λ) random variables, find the bias correction factor c_4 for the sample standard deviation. The APPL statements

```
> X := ExponentialRV(lambda):
> Y := Difference(X, X):
> g := [[x -> -x, x -> x], [-infinity, 0, infinity]]:
> Z := Transform(Y, g):
> Mean(Z) / sqrt(2 * Variance(X)):
```

yield $c_4 = \sqrt{2}/2 \cong 0.707$.

APPL also successfully executes the same code for $n = 2$ for the normal distribution ($c_4 = \sqrt{2/\pi} \cong 0.798$, which matches the $n = 2$ tabulated value exactly), exponential distribution ($c_4 = \sqrt{2}/2 \cong 0.707$), Erlang distribution ($c_4 = 3/4$), hyperbolic secant distribution ($c_4 \cong 0.768$), Rayleigh distribution ($c_4 = \frac{\sqrt{2\pi}-\sqrt{\pi}}{\sqrt{4-\pi}} \cong 0.792$), and the $U(0,1)$ distribution ($c_4 = \sqrt{6}/3 \cong 0.816$).

When $n = 3$, the mean of the sample standard deviation is

$$E\left[S\right] = \frac{1}{\sqrt{6}}\, E\left[\sqrt{2X_1^2 + 2X_2^2 + 2X_3^2 - 2X_1X_2 - 2X_1X_3 - 2X_2X_3}\,\right].$$

The appearance of the random variables X_1, X_2, and X_3 at various positions in the expected value expression make the evaluation of $E\left[S\right]$ more difficult. Monte Carlo simulation must be relied on to provide the bias correction factors c_4 and c_5. Table 9.2 provides estimates of c_4 and c_5 using ten million replications (which ensures that the factors are accurate to three digits after the decimal point) for the same distributions considered in Table 9.1. The $n = 2$ row and normal columns are consistent with the exact results provided by APPL.

9.4 Conclusions

The control chart constants d_2, d_3, c_4, and c_5 can be calculated symbolically using a computer algebra system in the case of sampling from a normal population. In addition, d_2 and d_3 can be calculated symbolically for several

Table 9.2. Values of c_4 and c_5 for exponential, normal, Rayleigh, and $U(0,1)$ sampling distributions obtained by Monte Carlo simulation

	c_4				c_5			
n	Expon	Normal	Rayleigh	$U(0,1)$	Expon	Normal	Rayleigh	$U(0,1)$
2	0.707	0.798	0.792	0.816	0.707	0.602	0.610	0.577
3	0.797	0.886	0.882	0.912	0.604	0.463	0.472	0.410
4	0.839	0.920	0.917	0.946	0.544	0.393	0.398	0.324
5	0.865	0.938	0.935	0.962	0.501	0.346	0.354	0.272
6	0.883	0.949	0.948	0.972	0.469	0.314	0.318	0.237
7	0.897	0.957	0.956	0.977	0.443	0.289	0.294	0.212
8	0.907	0.963	0.964	0.981	0.420	0.270	0.267	0.194
9	0.916	0.967	0.967	0.984	0.401	0.254	0.254	0.180
10	0.923	0.971	0.969	0.986	0.386	0.240	0.245	0.169

non-normal populations and c_4 and c_5 can be calculated symbolically for several non-normal populations when $n = 2$. These calculations were performed with the aid of the Maple-based APPL software. Monte Carlo simulation can be used to estimate control chart constants that cannot be calculated symbolically.

One area of further study might exist in searching for relationships analogous to Eq. (9.7), which depend only on the single parameter n. Burr [24] presents a strong case for normal sampling applications; however, Wheeler [165] notes a trough in Burr's skewness versus kurtosis plot (especially in U-shaped distributions) where using normal-sampling-based control chart constants would severely misrepresent the population.

10

Linear Approximations of Probability Density Functions

Lee S. McDaniel, Andrew G. Glen, and Lawrence M. Leemis

Abstract We develop a method for approximating the PDF of a continuous random variable with a piecewise-linear function. Four algorithms for choosing the endpoints of the linear segments are compared. The approximation is applied to estimating the convolution of two independent random variables.

Keywords Convolutions • Optimization

10.1 Approximating a PDF

A PDF, which we denote by $f(x)$, may have a complex functional form leading to difficult computations. In particular, the convolution of some distributions may not yield a closed-form solution. Therefore, we seek an approximation

---

This original paper uses many aspects of APPL in order to make approximations to certain random variable algebra operations that do not have closed-form solutions. By making linear approximations of PDFs, APPL can then do operations such a `Convolution`, `Product`, and `Transform` in order to approximate probability distribution of the new random variable. APPL's ability to define piecewise functions allows simpler linear piecewise functions to be used to approximate the true PDFs. Furthermore, the optimal placement of the segments can be embedded in the APPL explorations.

L.S. McDaniel
Loiusiana State University, Baton Rouge, LA, USA

A.G. Glen
Department of Mathematics and Computer Science, The Colorado College, Colorado Springs, CO, USA

L.M. Leemis (✉)
The College of William and Mary, Williamsburg, VA, USA
e-mail: leemis@math.wm.edu

A.G. Glen, L.M. Leemis (eds.), *Computational Probability Applications*, International Series in Operations Research & Management Science 247,
DOI 10.1007/978-3-319-43317-2_10

to $f(x)$ that has a simpler functional form. We construct an approximation, denoted $\hat{f}(x)$, that is composed of piecewise-linear segments and is therefore easier to manipulate. To illustrate, Figure 10.1 is a plot of the PDF of a beta$(2, 4)$ random variable with $\hat{f}(x)$ constructed using $n = 10$ endpoints (which define nine line segments).

Given that we wish to approximate $f(x)$ with $n - 1$ piecewise linear segments, we must decide where the n endpoints of our segments, $x_1, x_2, \ldots, x_n$, should be placed. There are many sensible methods for placing the endpoints.

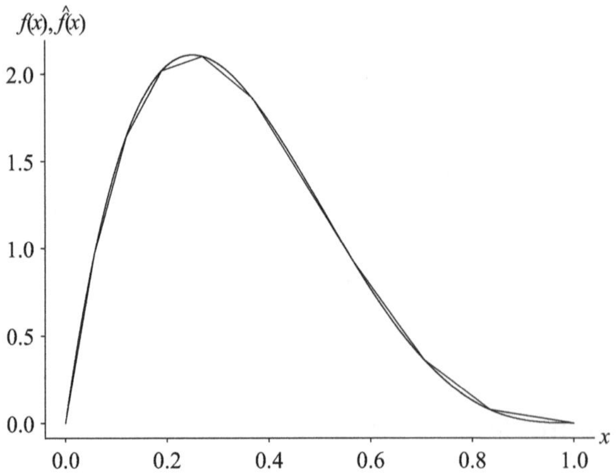

Fig. 10.1. A piecewise linear approximation to the beta$(2, 4)$ distribution with $n = 10$

The *Mathematica* function `ListInterpolation` can be used to approximate a function, but relies on the user to provide the endpoints, or simply defaults to equal spacing [171]. Interpolation using higher-order polynomials typically seeks optimal Lebesgue constants and is discussed by Brutman [26] and Smith [148]. Nguyen et al. [122] describe a method for approximation of parameterized functions that employs optimization techniques along with a rejection of a generalized basis set (such as polynomials) in favor of a problem-specific basis set. Rivlin [136] provides an overview of function approximation using various techniques.

In order to decide between the methods we present, a quantity must be defined that measures the quality of the fit. Given two piecewise-linear PDFs that approximate $f(x)$ with n endpoints, we select the approximation with the smallest L^2 distance from $f(x)$. So our metric is

$$D = D(x_1, x_2, \ldots, x_n) = \sqrt{\int_{-\infty}^{\infty} \left(\hat{f}(x) - f(x)\right)^2 dx}.$$

The method is organized as follows. In Section 10.2, we describe four methods for endpoint placement. This is followed by a comparison of the methods for several popular parametric distributions in Section 10.3. Section 10.4 contains an application based on the convolution of independent random variables. Section 10.5 contains conclusions.

10.2 Methods for Endpoint Placement

With the metric in mind, we must decide where to place the endpoints of the line segments in order to obtain the best approximation to $f(x)$. This section describes four methods for choosing $x_1, x_2, \ldots, x_n$ to minimize D. Note that each of these methods requires that x_1 is equal to the lower bound of the support, and x_n is equal to the upper bound of the support. In the case of a support that is unbounded in both directions, we arbitrarily set x_1 to the 1st percentile of the distribution and set x_n to the 99th percentile. If the distribution is not bounded above (below), we define upper (lower) bound of the support as the 99th (1st) percentile of the original distribution. Additionally, because these approximations will not generally integrate to one, we normalize the piecewise linear approximation so that it integrates to one.

10.2.1 Equal Spacing

The simplest rule for endpoint placement is to evenly space them along the support of the distribution. For instance, if we seek to approximate a beta distribution, which has its support on the interval $[0, 1]$, with four line segments ($n = 5$), the endpoints are

$$x_1 = 0, x_2 = 1/4, x_3 = 1/2, x_4 = 3/4, x_5 = 1.$$

This is a very fast method from a computational standpoint, but does not take into account any information about the PDF other than its support. Our next method attempts to utilize some information about the distribution in an attempt to improve the quality of the approximation.

10.2.2 Placement by Percentiles

Another simple method for point placement employs equally-spaced percentiles. If, for example, we want to use $n = 5$ endpoints to approximate a distribution with finite support, we place points at the minimum and maximum of the support, as well as the 25th, 50th, and 75th percentiles. Figure 10.2 shows how to place endpoints using percentiles with $n = 10$ points yielding an approximation to the beta$(2, 4)$ distribution.

This method ignores the shape of $f(x)$, but takes into account some information about what may be considered the important aspects of $f(x)$. We next introduce a method that uses information about the shape of $f(x)$ to place points.

10.2.3 Curvature-Based Approach

In order to take the shape of $f(x)$ into account, we examine the curvature of the distribution. If $y = f(x)$, the curvature of $f(x)$ is given by Stewart [154]

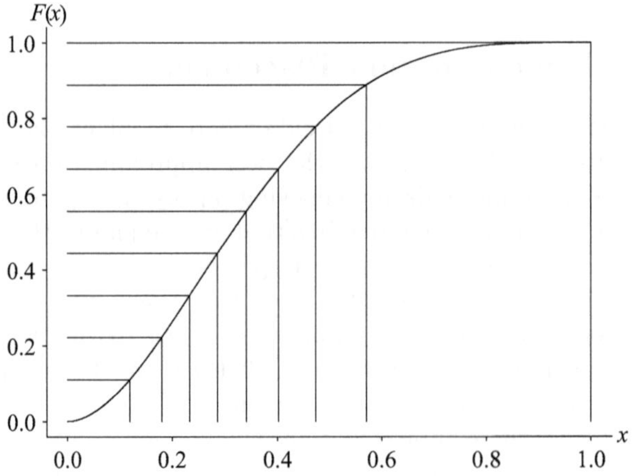

Fig. 10.2. Selection of endpoints using equally spaced percentiles

$$\kappa(x) = \frac{|y''|}{\left(1 + (y')^2\right)^{3/2}}.$$

This function quantifies the non linearity of $f(x)$. The greater the magnitude of the curvature over an interval, the more the function deviates from linearity on that interval. This suggests that more points should be placed in intervals for which the curvature has a higher magnitude.

The curvature function is not typically easy to manipulate. Therefore, we must evaluate the function at several points over the support of $f(x)$. Additionally, we want to find the points at which the function deviates from linearity the most without regard to the direction of the concavity. To address these concerns, we simply discretize the cumulative curvature function, which we define as

$$K(x) = \int_{-\infty}^{x} \kappa(x)dx = \int_{-\infty}^{x} \frac{|y''|}{\left(1 + (y')^2\right)^{3/2}} dx.$$

If we apply the method of placement by percentiles to this cumulative curvature function instead of $F(x)$, we get points that are placed more densely along the support of $f(x)$ when the function deviates more from linearity. Because $K(x)$ tends to increase very quickly, and therefore cluster too many endpoints too tightly, we need to take logarithms of $\kappa(x)$ to spread the points out. So we instead apply the method of placement by percentiles to $\tilde{K}(x)$, which we define as

$$\tilde{K}(x) = \int_{-\infty}^{x} \ln\left(\ln\left(\kappa(x)+1\right)+1\right) dx.$$

Figure 10.3 shows how we place our points when we consider a beta(2, 4) distribution with $n = 10$, while Figure 10.4 shows where the points are actually placed on $f(x)$. For this distribution, the curvature is given by

$$\kappa(x) = \frac{\left|-240x^2 + 360x - 120\right|}{\left(1 + 400\left(-4x^3 + 9x^2 - 6x + 1\right)^2\right)^{3/2}}$$

for $0 < x < 1$. Then $\tilde{K}(x)$ evaluated at $x = 1$ is

$$\tilde{K}(1) \cong 0.466.$$

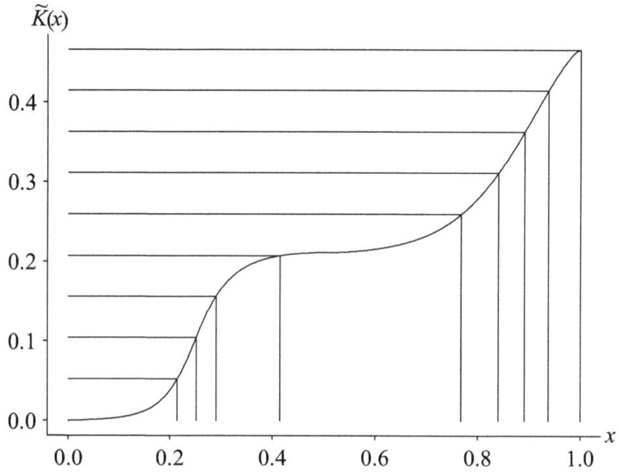

Fig. 10.3. Selection of endpoints using curvature

10.2.4 Optimization-Based Approach

Our fourth method is based on a mathematical program designed to minimize D. Given a PDF $f(x)$ with support $[a, b]$, we wish to solve the following constrained optimization problem

$$\begin{aligned}
\min_{x_1, x_2, \ldots, x_n} \quad & D(x_1, x_2, \ldots, x_n) \\
\text{s.t.} \quad & x_1 = a \\
& x_n = b \\
& \int_{-\infty}^{\infty} \hat{f}(x)dx = 1 \\
& x_1 < x_2 < \cdots < x_n.
\end{aligned}$$

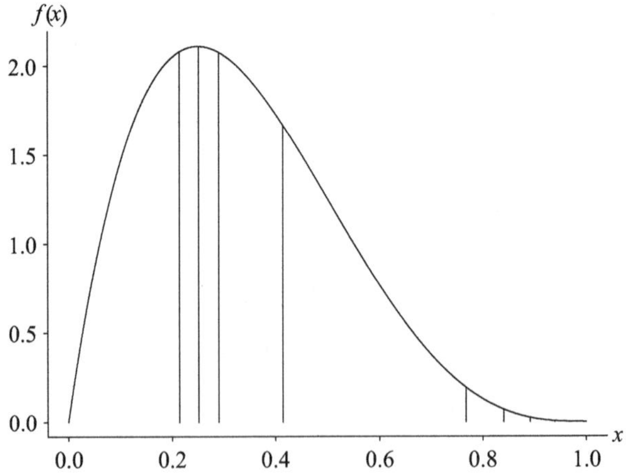

Fig. 10.4. Endpoint placement on $f(x)$

However, we prefer to avoid the use of an optimization package to solve this problem, so we must come up with a similar formulation that does not require an objective function. In order to make this a more solvable problem, we first remove the constraint requiring that $\hat{f}(x)$ integrates to one. We simply correct our $\hat{f}(x)$ after we find appropriate endpoints so that it integrates to one. We then note that this problem can be broken into pieces according to each line segment and that minimizing D yields the same solution as minimizing D^2. Now our problem becomes

$$\begin{array}{lll}
\displaystyle\min_{x_1,x_2,\ldots,x_n} & D(x_1,x_2)^2 + D(x_2,x_3)^2 + \cdots + D(x_{n-1},x_n)^2 & \\
\text{s.t.} & \int_{x_1}^{x_2} \left(\hat{f}(x) - f(x)\right)^2 dx & = D(x_1,x_2)^2 \\
& \int_{x_2}^{x_3} \left(\hat{f}(x) - f(x)\right)^2 dx & = D(x_2,x_3)^2 \\
& \vdots & \vdots \\
& \int_{x_{n-1}}^{x_n} \left(\hat{f}(x) - f(x)\right)^2 dx & = D(x_{n-1},x_n)^2 \\
& x_1 & = a \\
& x_n & = b \\
& x_1 < x_2 < \cdots < x_n, &
\end{array}$$

where $D(x_i, x_{i+1})^2$ is defined as the square of the L^2 distance between $f(x)$ and the line segment passing through the points $\left(x_i, f(x_i)\right)$ and $\left(x_{i+1}, f(x_{i+1})\right)$ on the interval $\left[x_i, x_{i+1}\right]$.

This modified formulation has n variables of the form x_i and $n-1$ variables of the form $D(x_i, x_{i+1})^2$, accompanied by $n+1$ equations. If we then constrain $D(x_1,x_2)^2 = D(x_2,x_3)^2 = \cdots = D(x_{n-1},x_n)^2 = C$, we have reduced our problem to

$$\begin{aligned}
\min_{x_1,x_2,\ldots,x_n} \quad & C(n-1) \\
\text{s.t.} \quad & \int_{x_1}^{x_2} \left(\hat{f}(x) - f(x)\right)^2 dx &= C \\
& \int_{x_2}^{x_3} \left(\hat{f}(x) - f(x)\right)^2 dx &= C \\
& \qquad \vdots & \vdots \\
& \int_{x_{n-1}}^{x_n} \left(\hat{f}(x) - f(x)\right)^2 dx &= C \\
& x_1 &= a \\
& x_n &= b.
\end{aligned}$$

We now have $n+1$ equations and $n+1$ unknowns, potentially giving us a unique feasible solution. We no longer need to rely on an optimization package, which makes the problem much easier to solve with a variety of different tools. This set of equations is solved numerically using equally-spaced endpoints as the initial estimate. Our solution for $x_1, x_2, \ldots, x_n$ using this heuristic will serve as our endpoints for $\hat{f}(x)$.

10.3 Comparison of the Methods

In addition to the metric D defined above, it is useful to plot the approximate PDFs to observe where each method succeeds and fails. For the beta$(2, 4)$ distribution, plots of the approximate PDFs generated by all four methods are shown in Figure 10.5 for $n = 10$. As expected the optimization-based method yields the smallest value for D, and appears to conform to the shape of the original distribution better than the other approximations.

To see how the methods compare for a range of other problems, we test various distributions and display the resulting values for D in Table 10.1. The column with the smallest value of D is set in boldface type.

The optimization-based method displays a clear advantage overall, with no method performing better on any test case. The method of percentiles tends to perform the worst, as it seldom provides a D value that is close to the best. In the next section, we present an example that employs the approximate PDF.

10.4 Application

Suppose we desire the value of the 80th percentile of the convolution of a beta$(9/4, 17/4)$ random variable and a beta$(21/4, 9/4)$ random variable, as well as the PDF of the convolution. Assuming independence, we present three techniques for finding our solutions. The convolution theorem is the first technique, and may supply us with the exact solution. Monte Carlo simulation is the second technique and is appealing for the 80th percentile, but is not well suited for finding the PDF of the convolution. Our third technique is to apply the convolution theorem to two approximate PDFs. Let $Z = X + Y$, where X and Y are independent random variables with $X \sim$ beta$(9/4, 17/4)$ and $Y \sim$ beta$(21/4, 9/4)$.

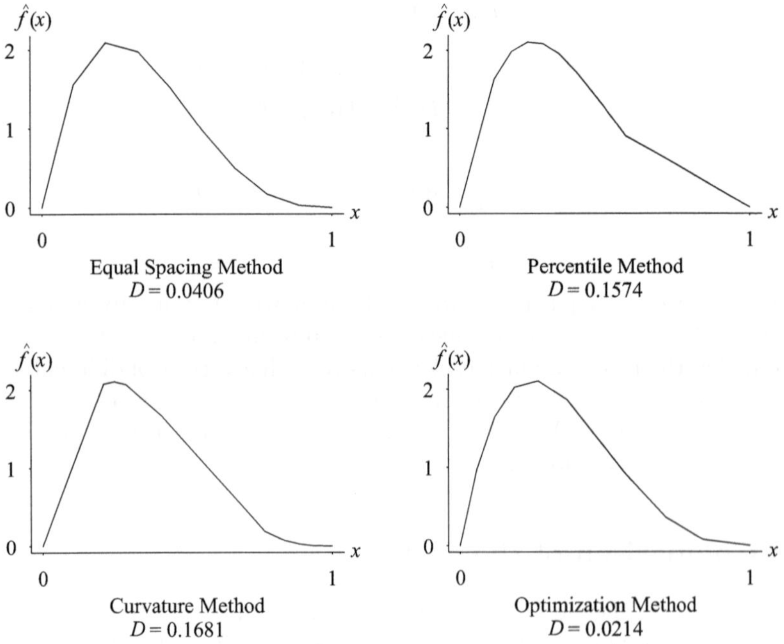

Fig. 10.5. Approximations to the beta(2, 4) distribution

10.4.1 Convolution Theorem

The convolution theorem gives an analytic method for finding the exact distribution of the convolution of two independent random variables. The theorem states that

$$f_Z(z) = \int_{-\infty}^{\infty} f_X(y - z) f_Y(z) dz.$$

The APPL code

```
> X := BetaRV(9 / 4, 17 / 4);
> Y := BetaRV(21 / 4, 9 / 4);
> Z := Convolution(X, Y);
```

fails to yield the exact distribution of the convolution in closed form due to mathematical intractability. Therefore, we must proceed to approximate techniques.

10.4.2 Monte Carlo Approximation

The Monte Carlo approximation of the 80th percentile is easy to find. Using 50 samples of size 10,000, we get a 95 % approximate confidence interval on the 80th percentile of

Table 10.1. Comparison of approximation methods with $n = 10$

	Equal spacing	Percentiles	Curvature	Optimization
Standard normal	0.0202	0.0317	0.0209	**0.0195**
Beta(2, 4)	0.0406	0.1574	0.1681	**0.0214**
Beta(4, 2)	0.0406	0.1574	0.1681	**0.0214**
Chi square(4)	0.0340	0.0279	0.0744	**0.0087**
Chi square(10)	0.0109	0.0335	0.0177	**0.0076**
Chi square(20)	0.0089	0.0482	0.0105	**0.0061**
Exponential(1)	0.0121	0.0391	0.0190	**0.0090**
Weibull(1, 3)	0.0277	0.0840	0.0336	**0.0246**
Gamma(1, 2)	0.0481	0.0395	0.0443	**0.0124**
Gamma(1, 5)	0.0155	0.0473	0.0237	**0.0107**
Gamma(1, 10)	0.0126	0.0682	0.0148	**0.0086**
Standard Cauchy	0.3088	0.1953	0.2805	**0.0566**
Pareto(1, 2)	0.1849	0.1396	0.1244	**0.0227**
Logistic(1, 1)	0.0163	0.0372	0.0173	**0.0142**
Logistic(1/2, 1)	0.0115	0.0263	0.0128	**0.0101**
Logistic(2, 1)	0.0481	0.0675	**0.0468**	**0.0468**

$$1.2461 \pm 0.0009.$$

Thus, to two decimal places, we report the 80th percentile as 1.25. Using Monte Carlo simulation, we are able to plot the histogram which gives insight into the shape of the distribution of the convolution, but no functional form for the PDF. Figure 10.6 is the histogram based on a sample of size 100,000. We are unsatisfied with the histogram as an approximation to the PDF, so we proceed to the convolution of approximate PDFs.

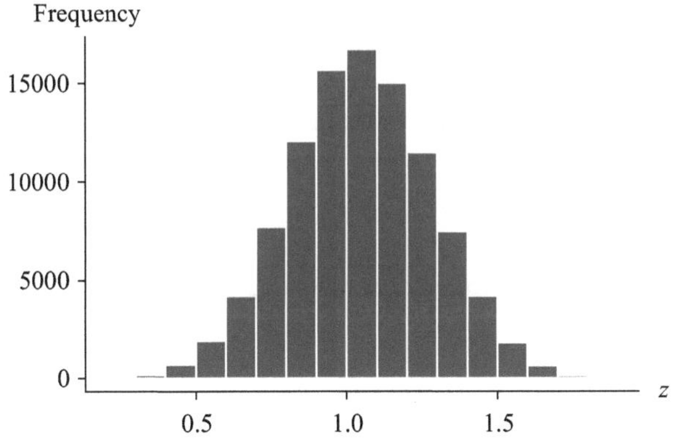

Fig. 10.6. Histogram of the convolution of X and Y

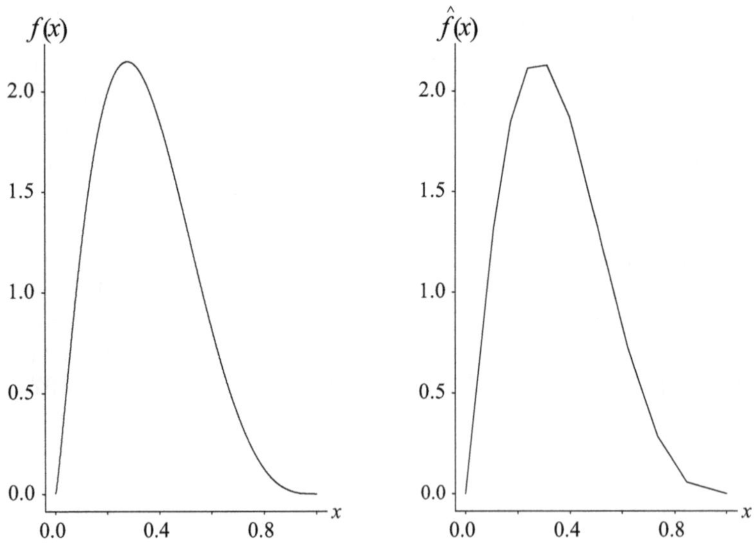

Fig. 10.7. The true and approximate distribution of X

10.4.3 Convolution of Approximate PDFs

Using the optimization-based method for selecting points, we find $\hat{f}_X(x)$ and $\hat{f}_Y(y)$. Figure 10.7 shows the true distribution and the approximated distribution of X with $n = 10$. We now apply the convolution theorem to $\hat{f}_X(x)$ and $\hat{f}_Y(y)$, each with $n = 10$, to find $\hat{f}_Z(z)$. This gives an estimated 80th percentile of 1.2478 which, to two decimal places, matches the value from Monte Carlo simulation. Figure 10.8 is a plot of the estimated distribution which exhibits the same shape as the histogram obtained via Monte Carlo simulation, but has the advantage of retaining continuity. This curve has $n^2 - 1 = 99$ segments. (If X and Y were independent and identically distributed, for example, there would be fewer segments in the support of $Z = X + Y$.) The functional form of the PDF is obviously too lengthy to display here. The APPL code to estimate the 80th percentile and PDF of Z is

```
> X := BetaRV(9 / 4, 17 / 4);
> Y := BetaRV(21 / 4, 9 / 4);
> Z := Convolution(ApproximateRV(X, 10), ApproximateRV(Y, 10));
> IDF(Z, 0.80);
> PlotDist(Z);
```

The code for `ApproximateRV` is listed in the appendix.

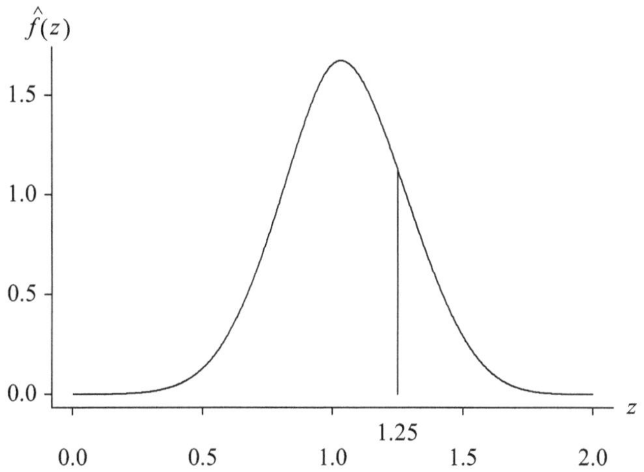

Fig. 10.8. Approximation of the convolution of X and Y and its 80th percentile

10.5 Conclusions

Functions of random variables, such as sums and products, can lead to problems that cannot be solved in closed form. These functions require approximation techniques that may yield unsatisfying results. We present a method for approximating a PDF with linear segments, which are easily manipulated, and four heuristics for endpoint placement: (1) equal spacing of points, (2) equally-spaced percentiles, (3) a curvature-based approach, and (4) an approach based on optimization of the L^2 distance. The optimization-based approach yields the best results in terms of the L^2 distance as well as approximating the shape of the original distribution.

Appendix: APPL Code for ApproximateRV

The following APPL code shows the complete algorithm for fitting linear PDF approximations to a continuous PDF.

```
#
#  ApproximateRV returns an approximate continuous pdf with
#                n - 1 segments
#  INPUTS:  OrigRV   -> a random variable in APPL format
#           npoints  -> any integer >= 3
#
#  OUTPUTS  ApproxRV -> the approx. random variable with npoints - 1
#                       linear segments approaching the original
#                       random variable
#
> ApproximateRV := proc(OrigRV :: list(list),
```

```
>   npoints :: {positive, integer})
> local i, n, nsegments, ApproxRV, leftend, rightend, supportsize, X,
>       ns, pdflist, newseg, area, fX, pdflist2, xlist, C, eqns,
>       vars, ans, j, range, tmpeqn, eqspc;
> if (nargs <> 2) then
>   print(`ERROR(ApproximateRV): This proc requires 2 arguments`):
>   RETURN():
> fi:
> nsegments := npoints - 1;
> if (nsegments < 2) then
>   print(`ERROR(ApproximateRV): Second parameter must be > 2`):
>   RETURN():
> fi:
> X := PDF(OrigRV);
#
#  Calculate the new support
#
> if (X[2][1] = -infinity) then
>   leftend := IDF(X, 1 / 100)
> else
>   leftend := X[2][1]
> fi;
> supportsize := nops(X[2]);
> if (X[2][supportsize] = infinity) then
>   rightend := IDF(X, 99 / 100)
> else
>   rightend := X[2][supportsize]
> fi;
#
#  Calculate equal spacing for initial solutions for x1, x2, ..., xn
#
> eqspc := [leftend];
> for i from 1 to nsegments - 1 do
>   eqspc := [op(eqspc), eqspc[i] + (rightend - leftend)
>            * (1 / (npoints - 1))];
> od;
> eqspc := [op(eqspc), rightend];
#
#  Formulate equations and format for fsolve
#
> eqns := [];
> xlist := [seq(xlist[i],i = 1 .. npoints)];
> xlist[1] := leftend;
> vars := [];
> for i from 2 to npoints do
>   tmpeqn := int(((((X[1][1](xlist[i]) - X[1][1](xlist[i - 1]))) /
>            (xlist[i] - xlist[i - 1])) * (x - xlist[i - 1]) +
>            X[1][1](xlist[i - 1]) - X[1][1](x)) ^ 2,
>            x = xlist[i - 1] .. xlist[i]) = C;
```

```
>   eqns := [op(eqns), tmpeqn];
>   vars := [op(vars), xlist[i] = eqspc[i]];
> od;
> eqns := [op(eqns), xlist[npoints] = rightend];
> vars := [op(vars), C = 0.01];
> range := [];
> for i from 1 to nops(vars) do
>   if (i < nops(vars)) then
>     range := [op(range), lhs(vars[i]) = leftend .. rightend];
>   else
>     range := [op(range) , lhs(vars[i]) = 0..1];
>   fi;
> od;
> ans := [op(fsolve({op(eqns)}, {op(vars)}))];

> for i from 1 to nops(ans) do
>   for j from 1 to nops(ans) do
>     if (lhs(ans[j]) = xlist[i]) then
>       xlist[i] := rhs(ans[j]);
>     fi;
>     if (lhs(ans[j]) = C) then
>       C := rhs(ans[j]);
>     fi;
>   od;
> od;
> ns := xlist;
#
#  Calculate the segments of the PDF
#
> pdflist := [];
> for i from 1 to nsegments do
>   newseg := (PDF(X, ns[i + 1]) - PDF(X, ns[i]))
>            / (ns[i + 1] - ns[i]) *(x - ns[i])
>            + PDF(X, ns[i]);
>   pdflist := [op(pdflist), unapply(newseg, x)]
> od;
#
#  Normalize the PDF so that it integrates to 1
#
> area := 0;
> fX := [pdflist, ns, ["Continuous", "PDF"]];
> for i from 1 to nsegments do
>   area := area + evalf(int(fX[1][i](y),
>           y = fX[2][i] .. fX[2][i + 1])):
> od:
> pdflist2 := [];
> for i from 1 to nsegments do
>   newseg := fX[1][i](x) / area;
>   pdflist2 := [op(pdflist2), unapply(newseg, x)]
```

```
> od;
> ApproxRV := [pdflist2, ns, ["Continuous", "PDF"]];
> RETURN(ApproxRV);
> end:
```

11

Univariate Probability Distributions

Lawrence M. Leemis, Daniel J. Luckett, Austin G. Powell, and Peter E. Vermeer

Abstract We describe a web-based interactive graphic that can be used as a resource in introductory classes in mathematical statistics. This interactive graphic presents 76 common univariate distributions and gives details on (a) various features of the distribution such as the functional form of the PDF and CDF, graphs of the PDF for various parameter settings, and values of population moments; (b) properties that the distribution possesses, for example, linear combinations of independent random variables from a particular distribution family also belong to the same

Originally published in the *Journal of Statistics Education*, Volume 20, Number 3 in 2012, this paper illustrates the use of APPL throughout its foundation. So much is APPL ingrained in the background work of the paper, it was not even mentioned as a reference book. Like people who crunch numbers in Excel or type papers in Word, often these two important software packages do not get mentioned in the reference section of a report. This is our vision for APPL also, that it is a tool used by researchers as part of their everyday work environment. Included at the end of this paper are two subsequent sections that have expanded the original manuscript that show the APPL verifications used in computations associated with the binomial and exponential distributions. These sections are found in the website of the diagram and highlight the APPL commands used to verify the results of this paper.

L.M. Leemis (✉)
The College of William and Mary, Williamsburg, VA, USA
e-mail: leemis@math.wm.edu

D.J. Luckett
William and Mary, Williamsburg, VA, USA

A.G. Powell
Hilton, New York, NY, USA

P.E. Vermeer
United States Coast Guard, New York, NY, USA

A.G. Glen, L.M. Leemis (eds.), *Computational Probability Applications*, International Series in Operations Research & Management Science 247,
DOI 10.1007/978-3-319-43317-2_11

distribution family; and (c) relationships between the various distributions, including special cases, transformations, limiting distributions, and Bayesian relationships. The interactive graphic went on-line on 11/30/12 at the URL `www.math.wm.edu/~leemis/chart/UDR/UDR.html`. As of May 2016, the site has had 42,000 page views from 35,000 unique visitors.

Keywords Continuous distributions • Discrete distributions • Distribution properties • Limiting distributions • Special cases • Transformations • Univariate distributions

11.1 Introduction

Introductory textbooks in probability and statistics often introduce univariate probability distributions in separate sections, which obscures both an understanding of the relationships between distributions and the properties that many of them have in common. Leemis and McQueston [95] designed a figure to highlight these relationships and properties. Song and Chen [150] redesigned the figure in a matrix format, making it easier to locate distributions. This article contains a description of a website that contains an interactive graphic of the figure at `www.math.wm.edu/~leemis/chart/UDR/UDR.html`. The user can also find proofs of the relationships and properties at this website.

The figure in Leemis and McQueston [95] contains 76 probability distributions. Of these, 19 are discrete and 57 are continuous. Figure 11.1 contains a screenshot of the upper-left-hand corner of the interactive graphic. An alphabetical list of the discrete and continuous probability distributions is displayed on the left-hand side of the screen, along with a slider bar to scroll through the list. Just above this list are four buttons labeled CHART, ABOUT, LINKS, and CONTACT. The CHART button displays the chart that contains the distributions. The ABOUT button gives information concerning the chart. The LINKS button contains links that might be of interest to a visitor. One such link connects the user to plots of the coefficient of variation vs. the skewness and the skewness vs. the kurtosis for the various distributions included in the website. The CONTACT button gives contact information for the developers of the interactive graphic. There are two ways to highlight a probability distribution. The first is to hover over the name of the distribution on the list at the far left side of the screen, which brings the probability distribution into view on the chart. The second is to hover over the name of the distribution in the body of the chart. In Figure 11.1, the cursor is hovering over the Zipf distribution. When hovering over the Zipf distribution, in either of the two manners just described, five changes occur on the interactive graphic. First, the Zipf distribution is highlighted in blue (colors may vary depending on the browser). Second, the boxes associated with distributions connected to the Zipf distribution are highlighted in black. Third, the arrows and labels associated with outgoing relationships from the Zipf distribution are highlighted in blue. Fourth, the arrows and labels associated with incoming relationships to the Zipf distribution (there are no such relationships for the Zipf distribution) are highlighted in black. Fifth, the other distributions that are not connected to the Zipf distribution are lightened in intensity. The number of distributions that are highlighted is helpful to indicate which distributions are at the center of probability theory (for example, the normal, binomial, exponential, uniform, and chi square distributions) and which are at the periphery (for example, the Polya, generalized Pareto, and doubly non central F distributions). Finally, just to the northwest of the box for the Zipf distribution are a + and − button for zooming.

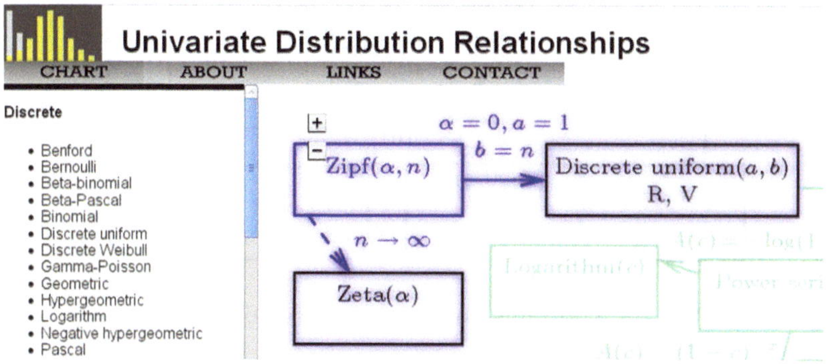

Fig. 11.1. Hovering over the Zipf distribution

Each box contains the distribution's name, its parameters, and a list of its properties using a single letter as a code (for example, C for the convolution property). For example, Figure 11.2 shows the interactive graphic when the cursor is hovering over the geometric distribution. The geometric distribution is highlighted in blue, while the Pascal (negative binomial) and discrete Weibull distributions are highlighted in black because they are connected to the geometric distribution. Since all three of these distributions are discrete, they are placed in rectangular boxes (the boxes for continuous distributions have rounded corners). The geometric distribution has one parameter, p, the probability of success. The Pascal distribution has two parameters, n, the number of successes, and p, the probability of success. The discrete Weibull distribution has two parameters, p and β. The geometric distribution has properties F (forgetfulness), M (minimum), and V (variate generation). Clicking the ABOUT tab reveals more detail about these properties. A legend in the lower-left-hand corner of the chart portion of the interactive graphic also contains the definitions of these properties. The Pascal distribution has the C (convolution) property, but only when the p parameter is fixed. The discrete Weibull distribution has the V (variate generation) property. More detail about the properties is given in Section 11.2. The interactive graphic also indicates how the two distributions are related by the arrows connecting the distributions. The cursor is hovering over the geometric distribution, so outgoing arrows are blue and the incoming arrows are black. The blue outgoing arrow indicates that the sum of mutually iid geometric (p) random variables has the Pascal distribution. This is known as a transformation. The incoming black arrows indicate that the geometric distribution is a special case of the Pascal distribution when $n = 1$ and a special case of the discrete Weibull distribution when $\beta = 1$. These are known as special cases. More detail about the relationships between distributions is given in Section 11.3.

Figure 11.3 shows the screen when the cursor hovers over the standard normal distribution. Solid arrows denote transformations and special cases.

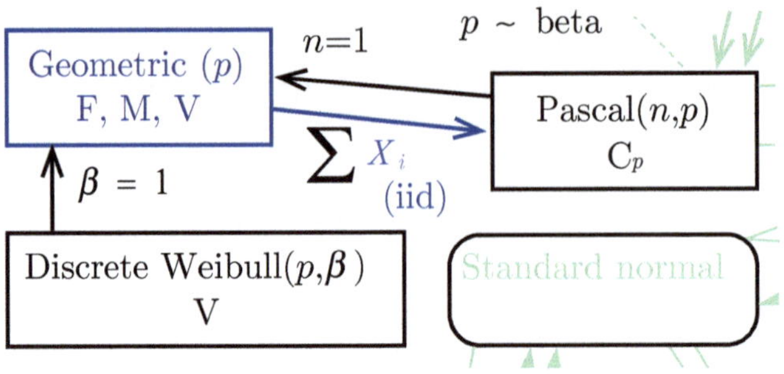

Fig. 11.2. Hovering over the geometric distribution

Dashed arrows denote asymptotic relationships, and dotted arrows denote Bayesian relationships. Some of the facts concerning the standard normal distribution from Figure 11.3 are listed below.

- The ratio of two independent standard normal random variables has the standard Cauchy distribution.
- The t distribution approaches the standard normal distribution as its degrees of freedom $n \to \infty$.
- The absolute value of a standard normal random variable has the chi distribution.
- The sum of squares of mutually independent standard normal random variables has the chi square distribution.
- The standard normal distribution is a special case of the normal distribution in which $\mu = 0$ and $\sigma = 1$.
- There is a one-to-one transformation between the normal distribution and standard normal distribution. Standardizing a normal random variable results in a standard normal random variable, which is useful for probability calculations. Multiplying a standard normal random variable by σ and adding μ results in a normal random variable, which is useful for random variate generation.

When the name of a distribution is clicked (either from the list at the far left or on the chart itself), a window appears or a download commences with a short description of the distribution. This description typically contains the PDF $f(x)$, the CDF $F(x)$, a graph of the PDF for various parameter values, the moment generating function, and the first four moments (the population mean, variance, skewness, and kurtosis). When these values are not mathematically tractable, they are excluded. In addition, when the cursor hovers over a property of a distribution, the associated letter turns red. Furthermore, when the user clicks on on the letter associated with a property of a distribution, a window opens with a proof of that particular property if a proof exists. For example, the URL for the minimum property of the geometric distribution

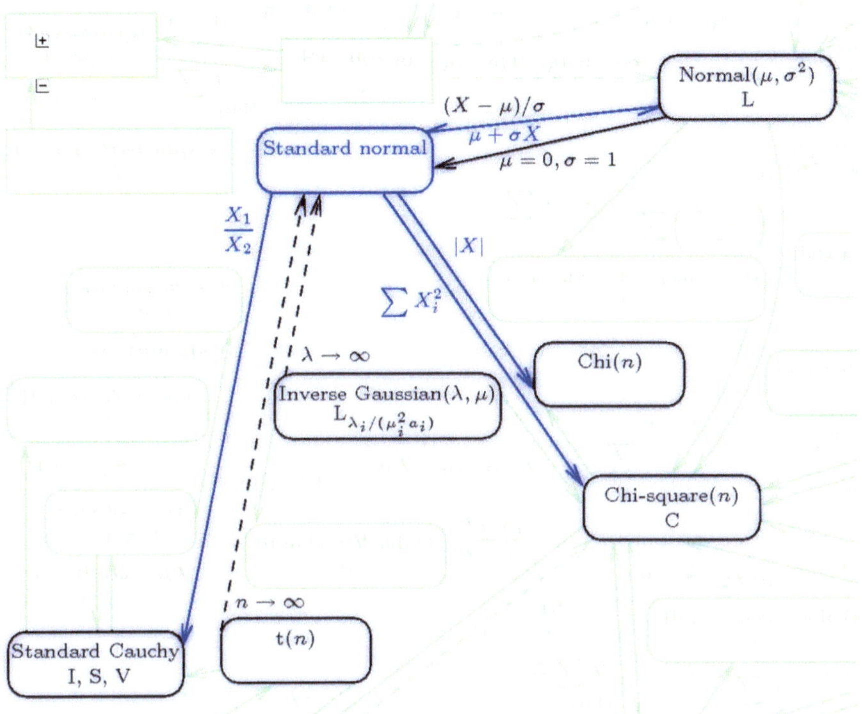

Fig. 11.3. Hovering over the standard normal distribution

is `http://www.math.wm.edu/~leemis/chart/UDR/PDFs/GeometricM.pdf`. Similar naming conventions are used for other distributions and their properties. An arrow turns red when the cursor is placed over the arrow. When an arrow is clicked, a window opens with a proof of the associated relationship if a proof exists. In some browsers, clicking on the backarrow button will return the user to the chart. In other browsers, simply closing the window returns the user to the chart. The authors were not able to prove the relationship, for example, between the inverse Gaussian distribution and the standard normal distribution. So when the associated arrow is clicked, a web page opens that cites the result and gives an unfinished proof. An invitation to the statistics community to complete any unfinished proofs and share the proofs with the developers for posting is hereby extended.

The interactive graphic is designed for practitioners who are interested in specific information about a distribution. It is also designed for instructors who are teaching courses in a mathematical statistics sequence. Listed here are several uses of classroom applications for the interactive graphic. First, in the initial probability class, typically fewer than a dozen discrete and continuous probability distributions are introduced. Exposing the students to the interactive graphic, however briefly, will let them know that there are a large number of more obscure univariate distributions that might arise in a modeling situation, many with well-developed theory. Second, the interactive graphic is an important reminder to the students that the univariate distributions are oftentimes related to one another. Third, when a proof of one of the properties or relationships is assigned as a homework exercise, the website can be used by students to see some of the techniques that are used to prove the relationships in the interactive graphic. Certain relationships between distributions were not included in the interactive graphic to keep the network of distributions planar. These "missing" proofs remain as viable homework exercises for students using the interactive graphic. The relationships can be between two distributions (for example, the floor of an exponential random variable has the geometric distribution), or combining two distributions to arrive at a third distribution (for example, the ratio of a standard normal random variable to the square root of an independent chi square random variable with n degrees of freedom divided by n is a t random variable with n degrees of freedom).

In Section 11.2 of this article we discuss the distribution properties shown in the interactive graphic. In Section 11.3, we discuss the relationships between distributions shown in the interactive graphic. Section 11.4 contains conclusions.

11.2 Discussion of Properties

In this section, we discuss a number of distribution properties. These properties apply to a single distribution, and are denoted with a capital letter in the online interactive graphic. Clicking on a property on the chart opens up a file containing a proof of the associated property that the distribution satisfies. In our discussion, the abbreviations used in the interactive graphic are listed in parentheses after the name of the property.

- The *Convolution* property (C) means that the sum of mutually independent random variables following this distribution belongs to the same family of distributions. In other words, if $X_1, X_2, \ldots, X_n$ are mutually independent random variables from a given distribution family with this property, then $\sum_{i=1}^{n} X_i$ belongs to the same distribution family.

- The *Forgetfulness* property (F), also known as the *memoryless* property, means that the conditional distribution of the random variable following this distribution is identical to the unconditional distribution. In other words, a random variable X that follows a distribution with the forgetfulness property satisfies $P(X > t + s | X > t) = P(X > s)$ for all values of s and t. The term forgetfulness is used in place of the more common "memorylessness" designation because the letter M was taken by the minimum property.
- The *Inverse* property (I) means that the inverse (reciprocal) of a random variable following this distribution family belongs to the same distribution family. In other words, if X belongs to a given distribution family with this property, then $1/X$ belongs to the same distribution family.
- The *Linear Combination* property (L) means that a linear combination of mutually independent random variables following this distribution belongs to the same family of distributions. In other words, if $X_1, X_2, \ldots, X_n$ are mutually independent random variables from a given distribution family with this property, and $a_1, a_2, \ldots, a_n$ are real-valued constants, then $\sum_{i=1}^{n} a_i X_i$ belongs to the same distribution family.
- The *Minimum* property (M) means that the minimum of n mutually independent random variables following this distribution belongs to the same family of distributions. In other words, if $X_1, X_2, \ldots, X_n$ are mutually independent random variables following a given distribution family with this property, then $X_{(1)}$ (the first order statistic) belongs to the same distribution family.
- The *Product* property (P) means that the product of mutually independent random variables from this distribution belongs to the same family of distributions. In other words, if $X_1, X_2, \ldots, X_n$ are mutually independent random variables from a given distribution family with this property, then $\prod_{i=1}^{n} X_i$ belongs to the same distribution family.
- The *Residual* property (R) means that the conditional distribution of a random variable from this distribution family that is left-truncated at a value in its support belongs to the same distribution family.
- The *Scaling* property (S) means that a random variable from this distribution family that is multiplied by a positive, real-valued constant belongs to the same distribution family. In other words, if X is a random variable following this distribution family, and a is a positive, real-valued constant, then aX belongs to the same distribution family.
- The *Variate Generation* property (V) means that the inverse of the CDF for this distribution family can be obtained in closed form. The inverse CDF can be used in a simple and fast algorithm to generate random variates from this distribution family for Monte Carlo and discrete-event simulation.

- The *Maximum* property (X) means that the maximum of mutually independent random variables from this distribution family belongs to the same distribution family. In other words, if $X_1, X_2, \ldots, X_n$ are mutually independent random variables from a given distribution family with this property, then $X_{(n)}$ (the nth order statistic) belongs to the same distribution family.

Note that property L implies properties C and S, and property F implies property R. These implications are listed in the legend in the lower-left-hand corner of the interactive graphic. We also note that certain properties only hold under restricted conditions. These properties are denoted on the chart with subscripts. For example, the binomial distribution is marked with C_p to show that the binomial distribution satisfies the *Convolution* property only when p is fixed.

11.3 Discussion of Relationships

There are many relationships between distributions. Some of the distributions that are included in the interactive graphic are generalizations of other distributions. For example, the exponential and chi square distributions are special cases of the gamma distribution. The Erlang distribution is the sum of mutually independent and identically distributed exponential random variables. The chi square distribution is the sum of squares of mutually independent standard normal random variables. The gamma distribution approaches the normal distribution as its shape parameter goes to infinity. We will now discuss each different type of relationship in more detail.

11.3.1 Special Cases

Many distributions are simply special cases of others. One well-known example is the standard normal distribution, which is a special case of the normal distribution where $\mu = 0$ and $\sigma^2 = 1$. In this example, the values of the parameters of the distribution are fully specified. There are other examples where only some of the parameters need to be specified. The exponential(α) distribution is a special case of the Weibull(α, β) distribution in which $\beta = 1$. In this case, α can remain unspecified.

11.3.2 Transformations

Other distributions are created through transformations. One way of distinguishing a transformation from a special case (since both are depicted by a solid arrow) is to recognize that a capital X will appear in the label by the arrow for a transformation. One of the most common transformations is to sum random variables. The binomial distribution with parameters n

and p, for example, is the sum of n mutually independent Bernoulli(p) random variables. The Pascal distribution with parameters n and p is the sum of n mutually independent geometric (p) random variables. In situations in which a sum is involved, the relationship between two distributions can be adequately described either with a transformation or a special case. For example, an Erlang(α, n) random variable is the sum of n mutually independent exponential(α) random variables. The exponential(α) distribution is a special case of the Erlang(α, n) distribution in which $n = 1$.

There are other transformations that are used to define random variables that do not involve a sum. For example, a chi random variable is the square root of a chi square random variable. A Rayleigh random variable is defined as the square of an exponential random variable. Transformations like these, which are expressed as one-to-one functions, can be inverted.

11.3.3 Limiting Distributions

A third type of relationship is a limiting distribution. This occurs when one distribution converges asymptotically to another distribution as one or more of the parameters approaches a limiting value. One important example is the gamma distribution, which converges to the normal distribution as the shape parameter β goes to infinity. This means that the gamma distribution can be used as a model when a normal distribution does not quite fit, or when an approximately normal distribution is needed, but only on positive support. Another important example is the t distribution, which approaches the standard normal distribution as the degrees of freedom, n, goes to infinity, which is widely used in statistical hypothesis tests.

Many non central models have a limiting distribution as the non centrality parameter approaches 0. For example, the non central F distribution converges to the F distribution as δ approaches 0. Finally, limiting distributions can give us insight into the behavior of the sum of random variables as the sample size becomes very large. The binomial, as the sum of mutually independent Bernoulli random variables, and the Erlang, as the sum of mutually independent exponential random variables, are examples of this. These both approach a normal distribution as $n \to \infty$ by the Central Limit Theorem.

11.3.4 Bayesian Models

A fourth type of relationship that is included on the chart is a stochastic parameter model. The familiar binomial model has parameters n and p, where n is a positive integer and $0 < p < 1$. Both parameters are assumed to be constants. But what of the case when one or both of the parameters are random variables? The dotted arrow that connects the binomial distribution to the beta-binomial distribution illustrates the case of allowing the parameter p to be a random variable having the beta distribution. The beta-Pascal, gamma-Poisson, and beta-binomial illustrate the particular type of relationship.

11.4 The Binomial Distribution

The following information is presented when one clicks on the binomial distribution bubble on the chart. Note the APPL commands at the end will produce the various properties of this distribution (as well as the others).

The shorthand $X \sim \text{binomial}(n,\, p)$ is used to indicate that the random variable X has the binomial distribution for positive integer parameter n and real parameter p satisfying $0 < p < 1$. The binomial distribution models the number of successes in n mutually independent Bernoulli trials, each with probability of success p. The random variable $X \sim \text{binomial}(n,\, p)$ has PDF

$$f(x) = \binom{n}{x} p^x \,(1-p)^{n-x} \qquad\qquad x = 0,\, 1,\, 2,\, \ldots,\, n.$$

The binomial distribution can be used to model the number of people in a group of n people with a particular characteristic, the number of defective items in a batch of n items, the number of fours in n rolls of a fair die, or the number of rainy days in a month. Stated more generically, a binomial random variable is the number of successes in n mutually independent Bernoulli trials. Three illustrations of the shape of the PDF for $n = 30$ and $p = 1/6,\, 1/2,\, 5/6$ are given in Figure 11.4.

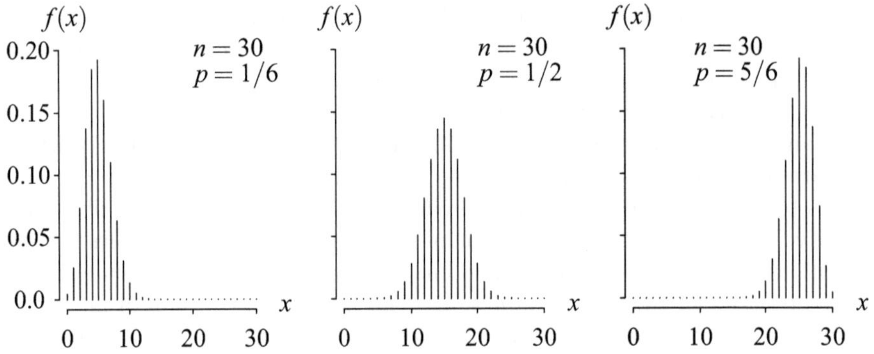

Fig. 11.4. Binomial PDFs for various values of n and p

The CDF on the support of X is

$$F(x) = P(X \le x) = \sum_{k=0}^{x} \binom{n}{k} p^k (1-p)^{n-k} \qquad\qquad x = 0,\, 1,\, 2,\, \ldots,\, n.$$

The survivor function on the support of X is

$$S(x) = P(X \geq x) = \sum_{k=x}^{n} \binom{n}{k} p^k (1-p)^{n-k} \qquad x = 0, 1, 2, \ldots, n.$$

The moment generating function of X is

$$M(t) = E\left[e^{tX}\right] = \left(1 - p + pe^t\right)^n \qquad -\infty < t < \infty.$$

The characteristic function of X is

$$\phi(t) = E\left[e^{itX}\right] = \left(1 - p + pe^{it}\right)^n \qquad -\infty < t < \infty.$$

The population mean and variance of a binomial(n, p) random variable are

$$E[X] = np \qquad \text{and} \qquad V[X] = np(1-p)$$

and the population skewness and kurtosis are

$$E\left[\left(\frac{X-\mu}{\sigma}\right)^3\right] = \frac{1-2p}{\sqrt{np(1-p)}} \qquad \text{and} \qquad E\left[\left(\frac{X-\mu}{\sigma}\right)^4\right] = 3 + \frac{1-6p(1-p)}{np(1-p)}.$$

The population skewness and kurtosis converge to 0 and 3, respectively, in the limit as $n \to \infty$.

Let $x_1, x_2, \ldots, x_n$ be realizations of mutually independent Bernoulli(p) random variables. Assume that n is a fixed constant and that p is an unknown parameter satisfying $0 < p < 1$. The maximum likelihood estimator for p is

$$\hat{p} = \frac{1}{n} \sum_{i=1}^{n} x_i,$$

which is an unbiased estimator of p, that is $E[\hat{p}] = p$. An approximate $(1-\alpha)100\,\%$ confidence interval for p is

$$\hat{p} - z_{\alpha/2}\sqrt{\frac{\hat{p}(1-\hat{p})}{n}} < p < \hat{p} + z_{\alpha/2}\sqrt{\frac{\hat{p}(1-\hat{p})}{n}},$$

where $z_{\alpha/2}$ is the $1-\alpha/2$ percentile of the standard normal distribution. This confidence interval is symmetric about $\hat{p}$ and allows for an upper limit that is greater than 1 and a lower limit that is less than 0. A second approximate $(1-\alpha)100\,\%$ confidence interval for p is

$$\frac{\hat{p} + \frac{z_{\alpha/2}^2}{2n} + z_{\alpha/2}\sqrt{\frac{\hat{p}(1-\hat{p})}{n} + \frac{z_{\alpha/2}^2}{4n^2}}}{1 + z_{\alpha/2}^2/n} < p < \frac{\hat{p} + \frac{z_{\alpha/2}^2}{2n} - z_{\alpha/2}\sqrt{\frac{\hat{p}(1-\hat{p})}{n} + \frac{z_{\alpha/2}^2}{4n^2}}}{1 + z_{\alpha/2}^2/n}.$$

A third approximate $(1-\alpha)100\,\%$ confidence interval for p that is based on the Poisson approximation to the binomial distribution is

$$\frac{1}{2n}\chi^2_{2y,\,1-\alpha/2} < p < \frac{1}{2n}\chi^2_{2(y+1),\,\alpha/2},$$

where $y = x_1 + x_2 + \cdots + x_n$ and $\chi^2_{q,\,\beta}$ is the $1-\beta$ percentile of a chi square distribution with q degrees of freedom. A fourth approximate $(1-\alpha)\cdot 100\,\%$ confidence interval for p is

$$\left(1+\frac{n-y+1}{yF_{2y,\,2(n-y+1),\,1-\alpha/2}}\right)^{-1} < p < \left(1+\frac{n-y}{(y+1)F_{2(y+1),\,2(n-y),\,\alpha/2}}\right)^{-1},$$

where $F_{q,\,r,\,\beta}$ is the $1-\beta$ percentile of an F random variable with q and r degrees of freedom.

APPL verification: The APPL statements

```
> X := BinomialRV(n, p);
> Mean(X);
> Variance(X);
> Skewness(X);
> Kurtosis(X);
> MGF(X);
```

verify the population mean, variance, skewness, kurtosis, and moment generating function.

11.5 The Exponential Distribution

The following is the information presented when you click on the exponential distribution in the chart. The shorthand $X \sim \text{exponential}(\alpha)$ is used to indicate that the random variable X has the exponential distribution with positive scale parameter α. The exponential distribution can be parameterized by its *mean* α with the PDF

$$f(x) = \frac{1}{\alpha}e^{-x/\alpha} \qquad\qquad x > 0,$$

for $\alpha > 0$. An exponential random variable X can also be parameterized by its *rate* λ via the PDF

$$f(x) = \lambda e^{-\lambda x} \qquad\qquad x > 0,$$

for $\lambda > 0$. When the second parameterization is used, the meaning of the rate parameter depends on the application (for example, failure rate for reliability, arrival rate or service rate for queueing, recidivism rate in criminal justice). The exponential distribution is used in reliability to model the lifetime of an object which, in a statistical sense, does not age (for example, a fuse or light bulb). This property is known as the memoryless property. The exponential distribution is the only continuous distribution that possesses this property.

The only discrete distribution with the memoryless property is the geometric distribution. The exponential distribution is used in queueing theory to model the times between customer arrivals and the service times. The exponential distribution is used in survival analysis to model the lifetime of an organism or the survival time after treatment. The PDF using the first parametrization with $\alpha = 0.5, 1, 2$ is shown in Figure 11.5 below.

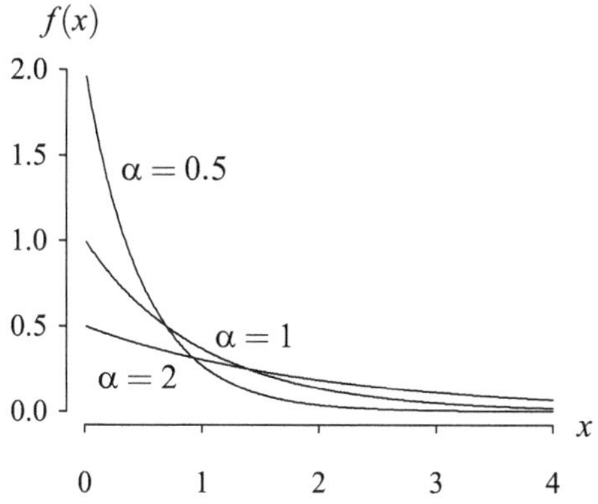

Fig. 11.5. Exponential PDFs for various values of α

Using the first parameterization, the CDF on the support of X is

$$F(x) = P(X \le x) = 1 - e^{-x/\alpha} \qquad x > 0.$$

The survivor function on the support of X is

$$S(x) = P(X \ge x) = e^{-x/\alpha} \qquad x > 0.$$

The hazard function on the support of X is

$$h(x) = \frac{f(x)}{S(x)} = \frac{1}{\alpha} \qquad x > 0.$$

The cumulative hazard function on the support of X is

$$H(x) = -\ln S(x) = \frac{x}{\alpha} \qquad x > 0.$$

The inverse distribution function of X is

$$F^{-1}(u) = -\alpha \ln(1-u) \qquad 0 < u < 1.$$

The median of X is

$$\alpha \ln 2.$$

The moment generating function of X is

$$M(t) = E\left[e^{tx}\right] = (1 - \alpha t)^{-1} \qquad t < \frac{1}{\alpha}.$$

The characteristic function of X is

$$\phi(t) = E\left[e^{itX}\right] = (1 - \alpha it)^{-1} \qquad t < \frac{1}{\alpha}.$$

The population mean, variance, skewness, and kurtosis of X are

$$E[X] = \alpha, \quad V[X] = \alpha^2, \quad E\left[\left(\frac{X-\mu}{\sigma}\right)^3\right] = 2, \quad E\left[\left(\frac{X-\mu}{\sigma}\right)^4\right] = 9.$$

For $X_1, X_2, \ldots, X_n$ mutually independent exponential(α) random variables, the maximum likelihood estimator for α is

$$\hat{\alpha} = \frac{\sum_{i=1}^{n} X_i}{n},$$

which is the sample mean. This is also the method of moments estimator.

APPL verification: The APPL statements

```
> assume(alpha > 0);
> X := [[x -> exp(-x / alpha) / alpha], [0, infinity],
>        ["Continuous", "PDF"]];
> CDF(X);
> SF(X);
> HF(X);
> CHF(X);
> Mean(X);
> Variance(X);
> Skewness(X);
> Kurtosis(X);
> MGF(X);
```

verify the CDF, survivor function, hazard function, cumulative hazard function, population mean, variance, skewness, kurtosis, and moment generating function.

11.6 Conclusions

The figure presented by Leemis and McQueston [95] is a helpful tool for students and instructors in the study of univariate probability distributions. It presents distributions simultaneously, as opposed to one at a time. It highlights how the distributions are related to each other, how distributions share many important properties, and how distributions are formed from other distributions. This figure is now available online as an interactive graphic. The online chart is more useful than the figure because it allows a user to click on distributions, properties, and relationships to display additional information, such as a proof, a graph, or various moments associated with a distribution.

12

Moment-Ratio Diagrams for Univariate Distributions

Erik Vargo, Raghu Pasupathy, and Lawrence M. Leemis

Abstract We present two moment-ratio diagrams along with guidance for their interpretation. The first moment-ratio diagram is a graph of skewness vs. kurtosis for common univariate probability distributions. The second moment-ratio diagram is a graph of coefficient of variation vs. skewness for common univariate probability distributions. Both of these diagrams, to our knowledge, are the most comprehensive to date. The diagrams serve four purposes: (1) they quantify the proximity between various univariate distributions based on their second, third, and fourth moments, (2) they illustrate the versatility of a particular distribution based on the range of values that the various moments can assume, (3) they can be used to create a short list of potential probability models based on a data set, and (4) they clarify the limiting

Originally published in the *Journal of Quality Technology*, Volume 42, Number 3, in 2010, this summary of moment-ratio diagrams is yet another example of how APPL worked behind the scenes to create diagrams that can be of use in probability modeling. In this case, the primary figures were built in a two-step process. First, every distribution's moments were calculated in APPL. In this step, the algorithm (in the appendix) was used to create the points on each moment curve for various parameter combinations of each distribution. This set of points was then imported into the statistical package R, which turned the points into curves in the figures in the paper.

E. Vargo
MITRE, McLean, VA, USA

R. Pasupathy
Purdue University, West Lafayette, IN, USA

L.M. Leemis (✉)
The College of William and Mary, Williamsburg, VA, USA
e-mail: leemis@math.wm.edu

A.G. Glen, L.M. Leemis (eds.), *Computational Probability Applications*, International Series in Operations Research & Management Science 247,
DOI 10.1007/978-3-319-43317-2_12

relationships between various well-known distribution families. The use of the moment-ratio diagrams for choosing a distribution that models given data is illustrated.

Keywords Coefficient of variation • Kurtosis • Skewness

12.1 Introduction

The *moment-ratio diagram* for a distribution refers to the locus of a pair of *standardized moments* plotted on a single set of coordinate axes (Kotz and Johnson [82]). By *standardized moments* we mean: the coefficient of variation (CV)

$$\gamma_2 = \frac{\sigma_X}{\mu_X},$$

the skewness (or third standardized moment)

$$\gamma_3 = E\left[\left(\frac{X-\mu_X}{\sigma_X}\right)^3\right],$$

and the kurtosis (or fourth standardized moment)

$$\gamma_4 = E\left[\left(\frac{X-\mu_X}{\sigma_X}\right)^4\right],$$

where μ_X and σ_X are the mean and the standard deviation of the implied (univariate) random variable X. The classical form of the moment-ratio diagram, plotted upside down, shows the third standardized moment γ_3 (or sometimes its square γ_3^2) plotted as abscissa and the fourth standardized moment γ_4 plotted as ordinate. The plot usually includes all possible pairs (γ_3, γ_4) that a distribution can attain. Since $\gamma_4 - \gamma_3^2 - 1 \geq 0$ (see [156, Exercise 3.19, page 121]), the moment-ratio diagram for a distribution occupies some subset of the shaded region shown in Figure 12.1.

Moment-ratio diagrams, apparently first introduced by Craig [36] and later popularized by Johnson et al. [71] especially through the plotting of multiple distributions on the same axes, have found enormous expediency among engineers and statisticians. The primary usefulness stems from the diagram's ability to provide a ready "snapshot" of the relative versatility of various distributions in terms of representing a range of shapes. Distributions occupying a greater proportion of the moment-ratio region are thought to be more versatile owing to the fact a larger subset of the allowable moment pairs can be modeled by the distribution. Accordingly, when faced with the problem of having to choose a distribution to model data, modelers often estimate the third and fourth standardized moments (along with their standard error estimates), and plot them on a moment-ratio diagram to get a sense of which distributions

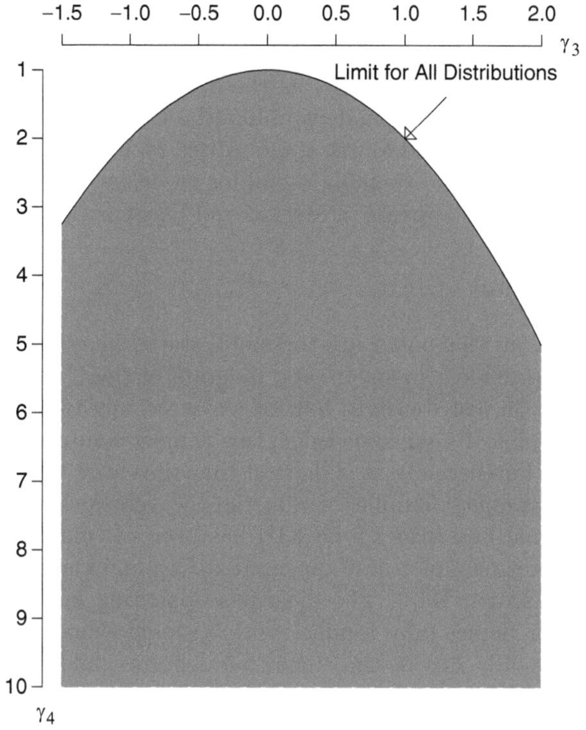

Fig. 12.1. The *shaded region* represents the set of attainable pairs of third and fourth standardized moments (γ_3, γ_4) for any distribution. The *solid line* is the limit $\gamma_4 = 1 + \gamma_3^2$ for all distributions

may be capable of representing the shapes implicit in the provided data. In this sense, a modeler can compare several "candidate" distributions simultaneously in terms of their moments. Another use for these diagrams has been in getting a sense of the limiting relationships between distributions, and also between various distributions within a system. An excellent example of the latter is the Pearson system of frequency curves where the region occupied by the various distributions comprising the system neatly divides the (γ_3^2, γ_4) plane (Johnson et al. [71, page 23]).

Since Craig [36] published the original moment-ratio diagram, various authors have expanded and published updated versions. The most popular of these happen to be the various diagrams appearing in Johnson et al. [71] (e.g., pages 23, 390). Rodriguez [138], in clarifying the region occupied by the Burr Type XII distribution in relation to others, provides a fairly comprehensive version of the moment-ratio diagram showing several important regions. Tadikamalla [158] provides a similar but limited version in clarifying the region occupied by the Burr Type III region.

More recently, Cox and Oakes [35] have popularized a moment-ratio diagram of a different kind—one that plots the CV (γ_2) as the abscissa and the third standardized moment (γ_3) as the ordinate. Admittedly, this variation is location and scale dependent unlike the classical moment-ratio diagrams involving the third and fourth standardized moments. Nevertheless, the diagram has become unquestionably useful for modelers. A slightly expanded version of this variation appears in Meeker and Escobar [113].

12.1.1 Contribution

Our contributions in this paper are threefold, stated here in order of importance. First, we provide a moment-ratio diagram of the CV versus skewness (γ_2, γ_3) involving 36 distributions, four of which occupy two-dimensional regions within the plot. To our knowledge, this is most comprehensive diagram available to date. Furthermore, it is the first time the entire region occupied by important two-parameter families within the CV versus skewness plot (e.g., generalized gamma, beta, Burr type XII) has been calculated and depicted. The CV versus skewness plot first appeared in Cox and Oakes [35], and later in Meeker and Escobar [113]. The diagrams appearing in both these original sources either depict only families with a single shape parameter (e.g., gamma), or vary only one of the shape parameters while fixing all others. Second, we provide a classical moment-ratio diagram (γ_3, γ_4) that includes 37 distributions, four of which occupy two-dimensional regions within the plot. While such diagrams are widely available, the diagram we provide is the most comprehensive among the sources we know, and seems particularly useful due to its depiction of all distributions in the same plot. In constructing the two moment-ratio diagrams, we have had to derive the limiting behavior of a number of distributions, some of which seem to be new. Expressions for γ_2, γ_3, and γ_4 for some of these distributions are listed in the appendix. We also host the moment-ratio diagrams in a publicly accessible website where particular regions of the diagram can magnified for clearer viewing. Third, using an actual data set, we demonstrate what a modeler might do when having to choose candidate distributions that "model" given data.

12.1.2 Organization

The rest of the paper is organized as follows. We present the two moment-ratio diagrams along with cues for interpretation in Sections 12.2, 12.3, and 12.4. Following that, we demonstrate the use of the moment-ratio diagrams for choosing a distribution that models given data in Section 12.5. Finally, we present conclusions and suggestions for further research in Section 12.6. This is followed by the appendix, where we provide analytical expressions for the moment-ratio locus corresponding to some of the distributions depicted in the diagrams, along with some of the APPL code that was used to create the diagrams.

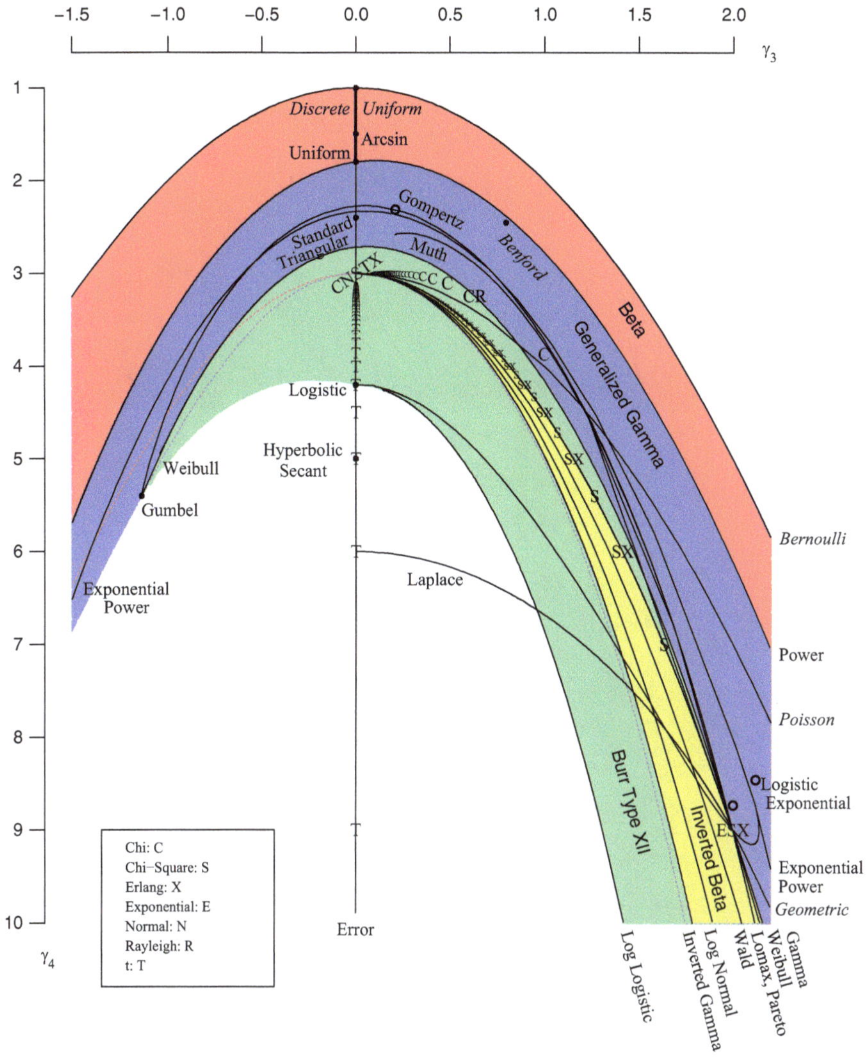

Fig. 12.2. Skewness (γ_3) versus Kurtosis (γ_4)

12.2 Reading the Moment-Ratio Diagrams

Two moment-ratio diagrams are presented in this paper. The first, shown in Figure 12.2, is a plot containing the (γ_3, γ_4) regions for 37 distributions. Figure 12.3 is a plot containing the (γ_2, γ_3) regions for 36 distributions.

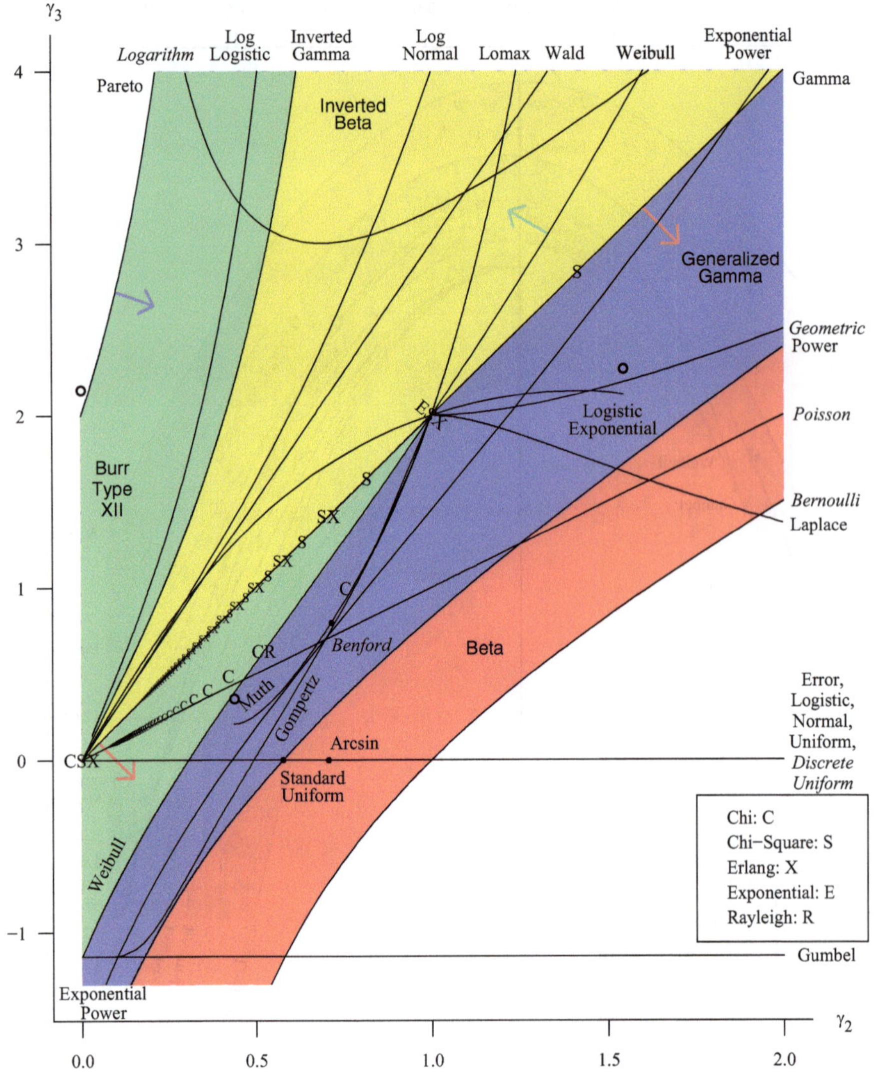

Fig. 12.3. CV (γ_2) versus Skewness (γ_3)

For convenience, in both diagrams, we have chosen to include discrete and continuous distributions on the same plot. In what follows we provide a common list of cues that will be useful in reading the diagrams correctly.

- Distributions whose moment-ratio regions correspond to *single points* (e.g., normal) are represented by black solid dots, *curves* (e.g., gamma) are represented by solid black lines, and *areas* (e.g., Burr Type XII) are represented by colored regions.
- The names of continuous distributions occupying a region are set in sans serif type; the names of continuous distributions occupying a point or curve are set in roman type; the names of discrete distributions occupying a point or curve are set in italics type.
- The end points of curves, when not attained by the distribution in question, are represented by an unfilled circle (e.g., logistic exponential).
- When the boundary of a moment-ratio area is obscured by another area, we include a dotted line (Figure 12.2) or an arrow (Figure 12.3) to clarify the location of the obscured boundary.
- When a distribution represented by points in one of the moment-ratio diagrams converges as one of its parameters approaches a limiting value (e.g., a t random variable as its degrees of freedom approaches infinity), we often decrease the font size of the labels to minimize interference.
- The parameterizations used for the distributions are from Leemis and McQueston [95] unless indicated otherwise in the paper.

12.3 The Skewness-Kurtosis Diagram

Whether the locus corresponding to a distribution in Figure 12.2 is a point, curve, or region usually depends on the number of shape parameters. For example, the normal distribution has no shape parameters and its locus in Figure 12.2 corresponds to the point $(0, 3)$. By contrast, since the gamma distribution has one shape parameter, its locus corresponds to the curve $\gamma_3 = 1.5\gamma_2^2+3$. An example of a distribution that has two shape parameters is the Burr Type XII distribution. It accordingly occupies an entire region in Figure 12.2. In all, Figure 12.2 has 37 distributions with 4 continuous distributions represented by regions, 19 distributions (15 continuous and 4 discrete) represented by curves, and 14 distributions (13 continuous and 1 discrete) represented by one or more points. A list of other useful facts relating to Figure 12.2 follows.

- The "T" plotted at $(\gamma_3, \gamma_4) = (0, 9)$ corresponds to the t distribution with five degrees of freedom, which is the smallest number of degrees of freedom in which the kurtosis exists.
- The chi square (S) and Erlang (X) distributions coincide when the chi square distribution has an even number of degrees of freedom. This accounts for the alternating pattern of "S" and "SX" labels that occur along the curve associated with the gamma distribution.
- Numerous distributions start at (or include) the locus of the normal distribution and end at (or include) the locus of the exponential distribution. Two examples of such are the gamma distribution and the inverted beta distribution.

- Space limitations prevented us from plotting the values associated with the discrete uniform distribution between its limits as a two-mass value with $(\gamma_3, \gamma_4) = (0, 1)$ and its limiting distribution (as the number of mass values increases) with $(\gamma_3, \gamma_4) = (0, 1.8)$. It is plotted as a thick line.
- The regime occupied by the inverted beta distribution has the curves corresponding to inverted gamma and the gamma distributions as limits.
- The regime occupied by the generalized gamma distribution has the curves corresponding to the power distribution and the log gamma distribution as partial limits.
- The regime occupied by the Burr Type XII distribution has the curve corresponding to the Weibull distribution as a partial limit.
- Barring extreme negative skewness values, virtually all of the regime occupied by the generalized gamma distribution is subsumed by the beta distribution.
- The beta and the Burr Type XII distributions seem complementary in the sense that the beta distribution occupies the "outer" regions of the diagram while the Burr Type XII distribution occupies the "inner" regions of the diagram. Furthermore, the collective regime of the beta and Burr Type XII distributions, with a few exceptions (e.g., Laplace), encompasses all other distributions included in the plot.

12.4 The CV-Skewness Diagram

Unlike in the skewness-kurtosis diagram (Figure 12.2), the locus of a distribution in the CV-skewness diagram (Figure 12.3) depends on the distribution's location and scale parameters. For this reason, in Figure 12.3, there are fewer distributions (compared to Figure 12.2) whose locus is a singleton. Figure 12.3 represents a total of 36 distributions with 4 continuous distributions represented by regions, 24 distributions (19 continuous and 5 discrete) represented by curves and 8 distributions (7 continuous and 1 discrete) represented by one or more points. A list of other useful facts relating to Figure 12.3 follows.

- Distributions that are symmetric about the mean have $\gamma_3 = 0$. Since CV can be adjusted to take any value (by controlling the location and scale), symmetric distributions, e.g., error, normal, uniform, logistic, have the locus $\gamma_3 = 0$ in Figure 12.3.
- The regime occupied by the beta family has the gamma curve $\gamma_3 = 2\gamma_2$, $\gamma_2 \in (0, 1)$ and the Bernoulli curve $\gamma_3 = \gamma_2 - 1/\gamma_2$ as limits.
- The regime occupied by the inverted beta distribution has the gamma curve $\gamma_3 = 2\gamma_2$, $\gamma_2 \in (0, 1)$ and the inverted gamma curve $\gamma_3 = 4\gamma_2 / \left(1 - \gamma_2^2\right), \gamma_2 \in (0, 1)$ as limits.
- The regime occupied by the generalized gamma distribution has the curves corresponding to the power distribution and the Pareto distribution as partial limits.

- The regime occupied by the Burr Type XII distribution has the curves corresponding to the Weibull and Pareto distributions as limits.

12.5 Application

The moment-ratio diagrams can be used to identify likely candidate distributions for a data set, particularly through a novel use of bootstrapping techniques, e.g., Cheng [31] and Ross [141]. Toward illustrating this, we first formally set up the problem. Let $X_1, X_2, \ldots, X_n$ be iid observations of a random variable having an unknown CDF $F(x)$. Suppose θ is some parameter concerning the population distribution (e.g., the coefficient of variation γ_2), and let $\hat{\theta}$ be its estimator (e.g., the sample coefficient of variation $\hat{\gamma}_2$ constructed from $X_1, X_2, \ldots, X_n$). Also let $F_n(x)$ denote the usual empirical CDF constructed from the data $X_1, X_2, \ldots, X_n$, i.e.,

$$F_n(x) = \frac{1}{n} \sum_{i=1}^{n} I\{X_i \le x\}.$$

A lot is known about how well $F_n(x)$ approximates $F(x)$. For example, the Glivenko–Cantelli theorem (from Billingsley [13]) states that $F_n \to F$ uniformly in x as $n \to \infty$. Furthermore, the deviation of $F_n(x)$ from $F(x)$ can be characterized fully through Sanov's theorem (from Dembo and Zeitouni [44]) under certain conditions.

We are now ready to demonstrate how the above can be used toward identifying candidate distributions to which a given set of data $X_1, X_2, \ldots, X_n$ might belong. As usual, the sample mean and sample standard deviation are calculated as

$$\bar{X} = \frac{1}{n} \sum_{i=1}^{n} X_i \qquad \text{and} \qquad S = \sqrt{\frac{1}{n-1} \sum_{i=1}^{n} \left(X_i - \bar{X}\right)^2}.$$

In order to obtain a nonzero standard deviation, we assume that at least two of the data values are distinct. The point estimates for the CV, skewness, and kurtosis are

$$\hat{\gamma}_2 = \frac{S}{\bar{X}}, \qquad \hat{\gamma}_3 = \frac{1}{n} \sum_{i=1}^{n} \left(\frac{X_i - \bar{X}}{S}\right)^3, \qquad \hat{\gamma}_4 = \frac{1}{n} \sum_{i=1}^{n} \left(\frac{X_i - \bar{X}}{S}\right)^4,$$

for $\bar{X} \neq 0$. The points $(\hat{\gamma}_3, \hat{\gamma}_4)$ and $(\hat{\gamma}_2, \hat{\gamma}_3)$ can be plotted in Figures 12.2 and 12.3 to give a modeler guidance concerning which distributions are potential parametric models for statistical inference. Probability distributions in the vicinity of the point estimates are strong candidates for probability models.

Unfortunately, these point estimates do not give the modeler a sense of their precision, so we develop an approximate interval estimate in the paragraph below.

Bootstrapping can be used to obtain a measure of the precision of the point estimates $(\hat{\gamma}_3, \hat{\gamma}_4)$ and $(\hat{\gamma}_2, \hat{\gamma}_3)$. Let B denote the number of bootstrap samples (a bootstrap sample consists of n observations drawn with replacement from the original data set). For each bootstrap sample, the two parameters of interest (e.g., skewness and kurtosis) are estimated using the procedure described in the previous paragraph and stored. After the B bootstrap samples have been calculated, the bivariate normal distribution is fitted to the B data pairs using standard techniques. Two of the five parameters of the bivariate normal distribution, namely, the two sample bootstrap means, are replaced by the point estimators to assure that the bivariate normal distribution is centered about the point estimators that were calculated and plotted in the previous paragraph. Finally, a concentration ellipse is plotted around the point estimate. The tilt associated with the concentration ellipse gives the modeler a sense of the correlation between the two parameters of interest.

Example 12.1. Consider the $n = 23$ deep-groove ball bearing failure times (measured in 10^6 revolutions)

17.88	28.92	33.00	41.52	42.12	45.60	48.48	51.84
51.96	54.12	55.56	67.80	68.64	68.64	68.88	84.12
93.12	98.64	105.12	105.84	127.92	128.04	173.40.	

from Lieblein and Zelen [101], which is discussed in Caroni [27]. For brevity, we consider the plotting of the point and associated concentration ellipse for only the CV vs. skewness moment ratio diagram (Figure 12.3). The first step is to calculate and plot the point $(\hat{\gamma}_2, \hat{\gamma}_3) \cong (0.519, 0.881)$. We then take $B = 200$ bootstrap samples of $n = 23$ failure times with replacement from the data set. (The value of B was chosen arbitrarily.) The bivariate normal distribution is fitted to the B data pairs and a concentration ellipse is then overlaid on the plot of the CV vs. skewness as a visual aid to identify likely candidate distributions for modeling the ball bearing lifetimes. The results of this process are displayed in Figure 12.4 which provides a close-up view of the concentration ellipse. In terms of candidate distributions, the following conclusions can be drawn.

- Because ball bearing lifetimes are inherently continuous, all of the discrete distributions should be eliminated from consideration.
- The position of the concentration ellipse implies that several distributions associated with regions in the (γ_2, γ_3) graph are candidate distributions: the gamma distribution (and its special cases), and the Weibull distribution (and the Rayleigh distribution as a special case) are likely to be models that fit the data well.
- The gamma and Weibull distributions both have shape parameters that are greater than 1 within the concentration ellipse, confirming

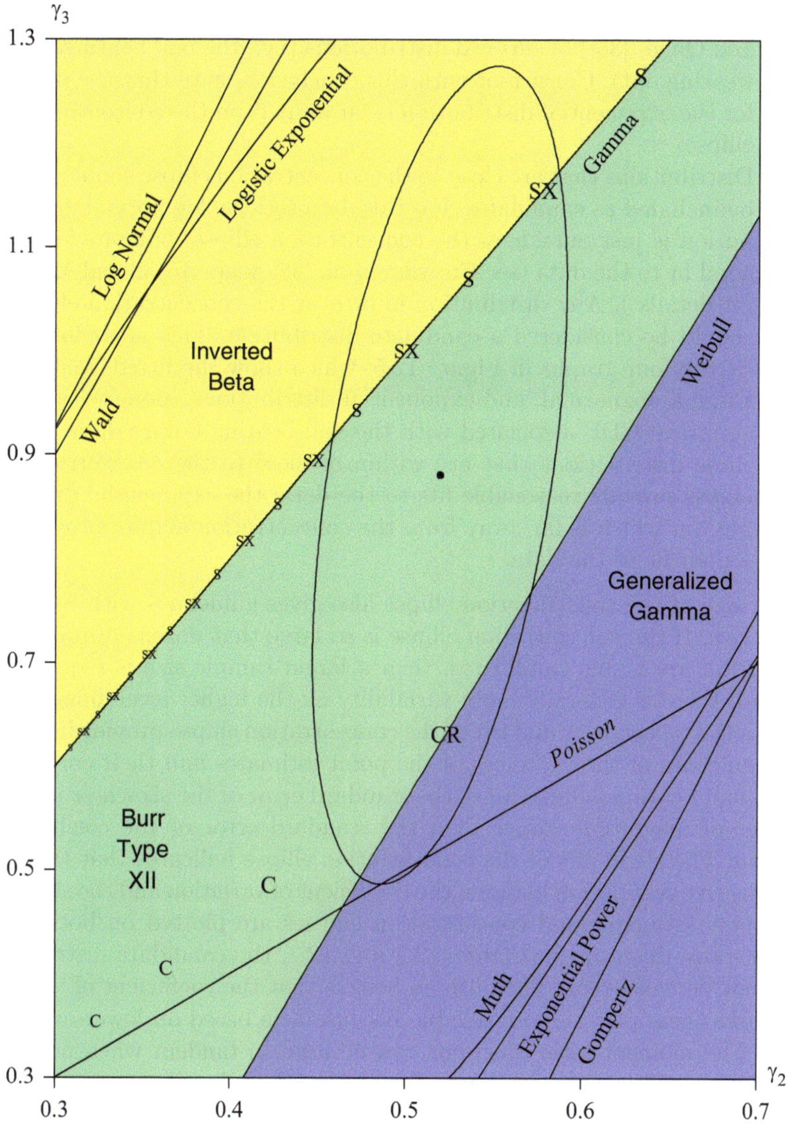

Fig. 12.4. Using the CV-skewness diagram to choose a candidate distribution (for modeling given data) through estimation and bootstrapping

the intuition that an appropriate model is in the IFR class (Cox and Oakes [35]) of survival distributions (i.e., the ball bearings are wearing out). Consistent with this conclusion, note that the point for the exponential distribution is far away from the concentration ellipse.

- Distributions that are close to the concentration ellipse should also be included as candidates. For this data set, the log normal distribution is just outside of the concentration ellipse, but provides a good fit to the data (see Crowder et al. [38] pages 37–38 and 42–43 for details). Any distribution in or near the concentration ellipse should be considered a candidate distribution. This is confirmed by the four graphs in Figure 12.5, which show the fitted Weibull, gamma, lognormal, and exponential distributions, along with the empirical CDF associated with the ball bearing failure data. The three distributions that are within or close to the concentration ellipse provide reasonable fits to the data; the exponential distribution, which is far away from the concentration ellipse, provides a poor fit to the data.

The size of the concentration ellipse also gives guidelines with respect to sample size. If the concentration ellipse is so large that dozens of probability distribution are viable candidates, then a larger sample size is required. As expected, there is generally more variability on the higher-level moments.

Also, the eccentricity and tilt of the concentration ellipse provide insight on the magnitudes of the variances of the point estimates and their correlation. For the ball bearing failure times, the standard error of the skewness is almost an order of magnitude larger than the standard error of the coefficient of variation. The slight tilt of the concentration ellipse indicates that there is a small positive correlation between the coefficient of variation and the skewness.

If point estimates and concentration ellipses are plotted on both of the moment-ratio diagrams in Figures 12.2 and 12.3, the candidate distributions might not be consistent. The authors believe that the coefficient of variation vs. the skewness plot is more reliable because it is based on lower-order moments. The moment-ratio diagrams can be used in tandem when using any one diagram still leaves a large number of candidate distributions.

12.6 Conclusions and Further Research

The two moment-ratio diagrams presented in Figures 12.2 and 12.3 are useful for insight concerning univariate probability distributions and for model discrimination for a particular data set. Plotting a concentration ellipse associated with bootstrap samples on either chart provides guidance concerning potential probability distributions that provide an adequate fit to a data set. These diagrams are one of the few ways that data analysts can simultaneously evaluate multiple univariate distributions.

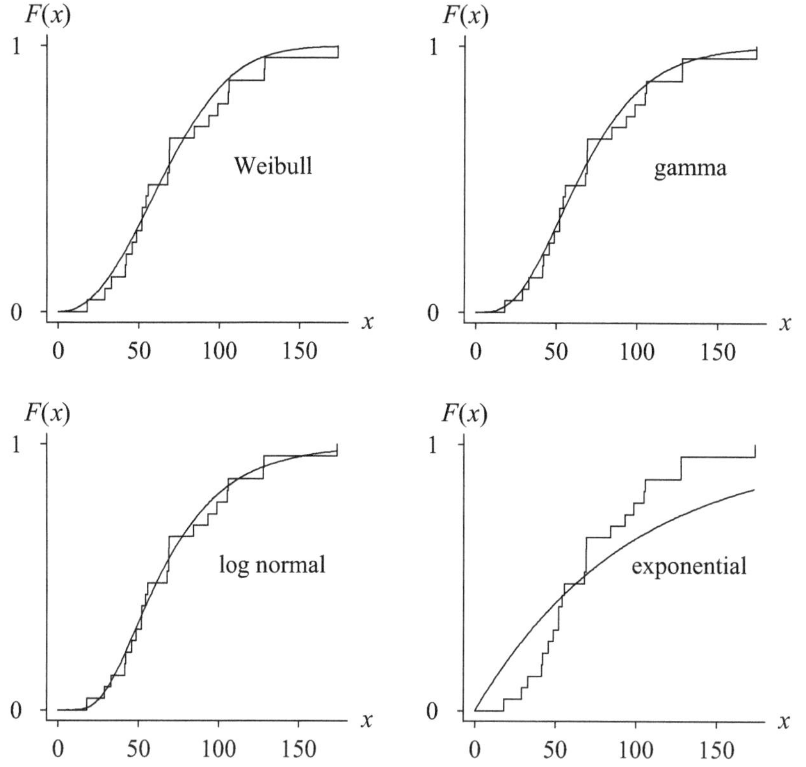

Fig. 12.5. Empirical and fitted CDFs for the Weibull, gamma, log normal, and exponential distributions for the ball bearing data set

Data sets associated with actuarial science, biostatistics, and reliability engineering often contain censored observations, with right-censored observations being the most common. Plotting the various moments is problematic for censored observations. Block and Leemis [16] provide techniques for overcoming censoring that are based on kernel density function estimation and competing risks theory. These techniques can be adapted to produce point estimators and concentration ellipses.

Further research work associated with these diagrams would include a Monte Carlo study that evaluates the effectiveness of the concentration ellipse in identifying candidate distributions. This study would indicate which of the two moment-ratio diagrams is better for model discrimination.

Appendix

In this section, we provide exact expressions for the CV, skewness, and kurtosis, for the four distributions that occupy (two-dimensional) regions in Figures 12.2 and 12.3.

Beta

The beta family (see Johnson et al. [72, Chap. 25, page 210]) has two shape parameters $p, q > 0$ with

$$\gamma_2 = \frac{\sqrt{q}}{\sqrt{p^2 + pq + p}}, \quad p, q > 0;$$

$$\gamma_3 = \frac{2(q-p)\sqrt{1/p + 1/q + 1/pq}}{p+q+2}, \quad p, q > 0;$$

$$\gamma_4 = 3(p+q+1)\frac{2(p+q)^2 + pq(p+q-6)}{pq(p+q+2)(p+q+3)}, \quad p, q > 0.$$

The regime in the (γ_2, γ_3) plane is bounded above by the line $\gamma_3 = 2\gamma_2$ corresponding to the gamma family, and below by the curve $\gamma_3 = \gamma_2 - 1/\gamma_2$. The regime in the (γ_3, γ_4) plane is bounded below by the limiting curve $\gamma_4 = 1 + \gamma_3^2$ for all distributions, and above by the curve $\gamma_4 = 3 + \frac{3}{2}\gamma_3^2$ corresponding to the gamma family.

Inverted Beta

The beta-prime or the Pearson Type VI family (see Johnson et al. [72, Chap. 25, page 248]), also known as the inverted beta family, has two shape parameters $\alpha, \beta > 0$ with

$$\gamma_2 = \sqrt{\frac{\alpha + \beta - 1}{\alpha(\beta - 2)}}, \quad \beta > 2;$$

$$\gamma_3 = \sqrt{\frac{4(\beta - 2)}{(\alpha + \beta - 1)\alpha}} \cdot \frac{2\alpha + \beta - 1}{\beta - 3}, \quad \beta > 3;$$

$$\gamma_4 = \frac{3(\alpha - 2 + \frac{1}{2}(\beta - 3)\gamma_2^2)}{\beta - 4}, \quad \beta > 4.$$

The regime in the (γ_2, γ_3) plane is bounded above by the curve $\gamma_3 = 4\gamma_2/(1 - \gamma_2^2), \gamma_2 \in (0, 1)$, and below by the curve $\gamma_3 = 2\gamma_2$ corresponding to the gamma family. The regime in the (γ_3, γ_4) plane is bounded above by the curve

$$\gamma_3 = \frac{4\sqrt{\alpha - 2}}{\alpha - 3}, \quad \gamma_4 = 3 + \frac{30\alpha - 66}{(\alpha - 3)(\alpha - 4)}, \quad \alpha > 4$$

corresponding to the inverse gamma family, and below by the curve $\gamma_4 = 3 + 3\gamma_3^2/2$ corresponding to the gamma family.

Generalized Gamma

The generalized gamma family (see Johnson et al. [71, page 388]) has two shape parameters $\alpha, \lambda > 0$ with the rth raw moment $\mu'_r = \Gamma(\alpha + r\lambda)/\Gamma(\alpha)$. The regime in the (γ_2, γ_3) plane is bounded below by the curve

$$\gamma_2 = \frac{1}{\sqrt{p(p+2)}}, \quad \gamma_3 = \frac{1-p}{p+3} \cdot \frac{2}{\sqrt{1+2/p}}, \quad p > 0$$

corresponding to the power family, and above by the curve

$$\gamma_2 = \frac{1}{\sqrt{p(p-2)}}, \quad \gamma_3 = \frac{1+p}{p-3} \cdot \frac{2}{\sqrt{1-2/p}}, \quad p > 3$$

corresponding to the Pareto family. The regime in the (γ_3, γ_4) plane is bounded above by the curve

$$\gamma_3 = \frac{1+p}{p-3} \frac{2}{\sqrt{1-2/p}}, \quad \gamma_4 = \frac{3(1+2/p)(3p^2-p+2)}{(p+3)(p+4)}, \quad p > 0$$

corresponding to the power family, bounded below to the right by the curve corresponding to the generalized gamma family with $\lambda = -0.54$, and bounded below to the left by the curve corresponding to the log gamma family. [Recall that the log gamma family with shape parameter $\alpha > 0$ has the rth cumulant $\kappa_r = \Psi^{(r)}(\alpha)$, where $\Psi^{(r)}(z)$ is the $(r+1)^{\text{th}}$ derivative of $\ln \Gamma(z)$.]

Burr Type XII

The Burr Type XII family (see Rodriques [138]) has two shape parameters $c, k > 0$ with the rth raw moment $\mu'_r = \Gamma(r/c+1)\Gamma(k-r/c)/\Gamma(k)$, $c > 0, k > 0, r < ck$. The regime in the (γ_2, γ_3) plane is bounded below by the curve corresponding to the Weibull family (r^{th} raw moment $\mu'_r = \Gamma(r/c+1)$, where $c > 0$ is the Weibull shape parameter), and above by the curve

$$\gamma_2 = \frac{1}{\sqrt{p(p-2)}}, \quad \gamma_3 = \frac{1+p}{p-3} \cdot \frac{2}{\sqrt{1-2/p}}, \quad p > 3$$

corresponding to the Pareto family. The regime in the (γ_3, γ_4) plane is bounded below by the curve corresponding to the Weibull family, bounded above to the right by the curve corresponding to the Burr Type XII family with $k = 1$, and bounded above to the left by the curve corresponding to the Burr Type XII family with $c = \infty$.

APPL Code for the Diagrams

The moment-ratio diagrams in this paper were created in two steps. For each distribution, an algorithm is used to create the sets of points that will produce the curves. These sets of points are then imported into R to produce the graphs. The APPL code to generate the points for the χ^2 distribution moment curves follow. Other distribution curves are produced similarly. Note the variable X establishes the χ^2_n distribution. The APPL commands in the fprintf statements find the desired moments of X for various values of n.

```
> file := fopen("ChiSquare.d", WRITE);
> n := 1;
> i := 1;
> while n < 30 do
>   X := ChiSquareRV(n);
>   fprintf(file, "%g %g %g %g\n", evalf(n), evalf(CoefOfVar(X)),
>           evalf(Skewness(X)),evalf(Kurtosis(X))):
>   i := i + 1:
>   n := i ^ 2:
> end do;
> for n to 100 do
>   X := ChiSquareRV(n);
>   if `mod`(n, 2) <> 0 then
>     fprintf(file, "%g %g %g %g\n", evalf(n), evalf(CoefOfVar(X)),
>             evalf(Skewness(X)), evalf(Kurtosis(X))):
>   end if
> end do:
> fprintf(file, "%g %g %g %g\n", evalf(999999), evalf(0), evalf(0),
>         evalf(3)):
> fclose(file):
```

13

The Distribution of the Kolmogorov–Smirnov, Cramer–von Mises, and Anderson–Darling Test Statistics for Exponential Populations with Estimated Parameters

Diane L. Evans, John H. Drew, and Lawrence M. Leemis

Abstract This paper presents a derivation of the distribution of the Kolmogorov–Smirnov, Cramer–von Mises, and Anderson–Darling test statistics in the case of exponential sampling when the parameters are unknown and estimated from sample data for small sample sizes via maximum likelihood.

Keywords Distribution functions • Goodness-of-fit tests • Maximum likelihood estimation • Order statistics • Transformation technique

Originally published in *Communications in Statistics—Simulation and Computation*, Volume 37, Number 7 in 2008, this paper contains a derivation of the probability distribution of some goodness of fit statistics when parameters are estimated from the data. It is possible in reality only with the environment of APPL to work on these unique distributions. Piecewise distributions like those in Figures 13.5 and 13.13 are one of the strengths of APPL analysis. Also the procedures `UniformRV` and `Transform` are used in calculating the distribution of the W_2^2 and A_2^2 statistics.

D.L. Evans
Rose Hulman, Terre Haute, IN, USA

J.H. Drew
William and Mary, Williamsburg, VA, USA

L.M. Leemis (✉)
The College of William and Mary, Williamsburg, VA, USA
e-mail: leemis@math.wm.edu

A.G. Glen, L.M. Leemis (eds.), *Computational Probability Applications*, International Series in Operations Research & Management Science 247, DOI 10.1007/978-3-319-43317-2_13

13.1 The Kolmogorov–Smirnov Test Statistic

The Kolmogorov–Smirnov (K–S) goodness-of-fit test compares a hypothetical or fitted CDF $\hat{F}(x)$ with an empirical CDF $F_n(x)$ in order to assess fit. The empirical CDF $F_n(x)$ is the proportion of the observations $X_1, X_2, \ldots, X_n$ that are less than or equal to x and is defined as

$$F_n(x) = \frac{I(x)}{n},$$

where n is the size of the random sample and $I(x)$ is the number of X_i's less than or equal to x.

The K–S test statistic D_n is the largest vertical distance between $F_n(x)$ and $\hat{F}(x)$ for all values of x, i.e.,

$$D_n = \sup_x \{|F_n(x) - \hat{F}(x)|\}.$$

The statistic D_n can be computed by calculating (see, for example, Law [88, page 364])

$$D_n^+ = \max_{i=1,2,\ldots,n} \left\{ \frac{i}{n} - \hat{F}(X_{(i)}) \right\}, \qquad D_n^- = \max_{i=1,2,\ldots,n} \left\{ \hat{F}(X_{(i)}) - \frac{i-1}{n} \right\},$$

where $X_{(i)}$ is the ith order statistic, and letting

$$D_n = \max\{D_n^+, D_n^-\}.$$

Although the test statistic D_n is easy to calculate, its distribution is mathematically intractable. Drew et al. [47] provide an algorithm for calculating the CDF of D_n when all the parameters of the hypothetical CDF $\hat{F}(x)$ are known (referred to as the all-parameters-known case). Assuming that $\hat{F}(x)$ is continuous, the distribution of D_n under H_0, where $X_1, X_2, \ldots, X_n$ are iid observations from a population with CDF $F(x)$, is a function of n, but does not depend on $F(x)$. Marsaglia et al. [106] provide a numerical algorithm for computing $\Pr(D_n \le d)$.

The more common and practical situation occurs when the parameters are unknown and are estimated from sample data, using an estimation technique such as maximum likelihood. In this case, the distribution of D_n depends upon both n and the particular distribution that is being fit to the data. Lilliefors [103] provides a table (obtained via Monte Carlo simulation) of selected percentiles of the K–S test statistic D_n for testing whether a set of observations is from an exponential population with unknown mean. Durbin [48] also provides a table (obtained by series expansions) of selected percentiles of the distribution of D_n. This paper presents the derivation of the *distribution* of D_n in the case of exponential sampling for $n = 1$, $n = 2$, and $n = 3$. Additionally, the distribution of the Cramer–von Mises and Anderson–Darling test statistics for $n = 1$ and $n = 2$ are derived in Section 13.2. Two case studies that analyze real-world data sets (Space shuttle accidents and commercial nuclear power accidents), where $n = 2$ and the fit to an exponential distribution is important, are given in Section 13.3. Future work involves extending the formulas established for the exponential distribution with samples of size $n = 1$, 2, and 3 to additional distributions and larger samples. For a summary of the literature available on these test statistics and goodness-of-fit techniques (including tabled values, comparative merits, and examples), see D'Agostino and Stephens [40].

We now define notation that will be used throughout this chapter. Let X be an exponential random variable with PDF $f(x) = \dfrac{1}{\theta}\, e^{-x/\theta}$ and CDF $F(x) = 1 - e^{-x/\theta}$ for $x > 0$ and fixed, unknown parameter $\theta > 0$. If $x_1, x_2, \ldots, x_n$ are the sample data values, then the maximum likelihood estimator (MLE) $\hat{\theta}$ is

$$\hat{\theta} = \frac{1}{n} \sum_{i=1}^{n} x_i.$$

We test the null hypothesis H_0 that $X_1, X_2, \ldots, X_n$ are iid exponential(θ) random variables.

13.1.1 Distribution of D_1 for Exponential Sampling

If there is only $n = 1$ sample data value, which we will call x_1, then $\hat{\theta} = x_1$. Therefore, the fitted CDF is

$$\hat{F}(x) = 1 - e^{-x/\hat{\theta}} = 1 - e^{-x/x_1} \qquad\qquad x > 0.$$

As shown in Figure 13.1, the largest vertical distance between the empirical CDF $F_1(x)$ and $\hat{F}(x)$ occurs at x_1 and has the value $1 - 1/e$, regardless of the value of x_1. Thus, the distribution of D_1 is degenerate at $1 - 1/e$ with CDF

$$F_{D_1}(d) = \begin{cases} 0 & d < 1 - 1/e \\ 1 & d \geq 1 - 1/e. \end{cases}$$

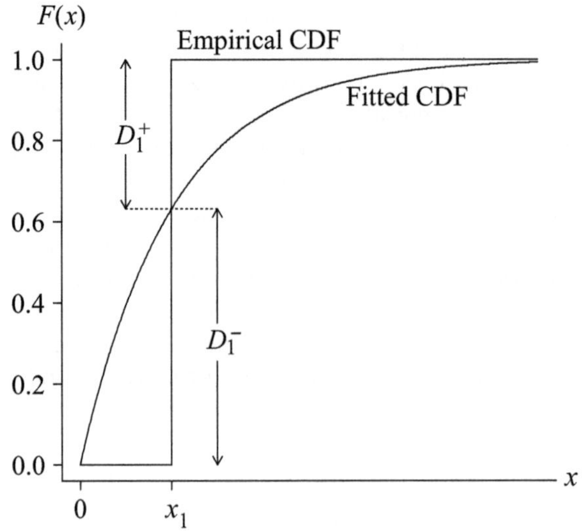

Fig. 13.1. The empirical and fitted exponential distribution for one data value x_1, where $D_1^- = 1 - 1/e$ and $D_1^+ = 1/e$. [Note: The riser of the empirical CDF in this and other figures has been included to aid in comparing the lengths of D_1^- and D_1^+]

13.1.2 Distribution of D_2 for Exponential Sampling

If there are $n = 2$ sample data values, then the MLE is $\hat{\theta} = (x_1 + x_2)/2$, and thus, the fitted CDF is

$$\hat{F}(x) = 1 - e^{-x/\hat{\theta}} = 1 - e^{-2x/(x_1+x_2)} \qquad x > 0.$$

A maximal scale invariant statistic (see Lehmann [99, page 215]) is

$$\frac{x_1}{x_1 + x_2}.$$

The associated statistic which is invariant to re-ordering is

$$y = \frac{x_{(1)}}{x_{(1)} + x_{(2)}},$$

where $x_{(1)} = \min\{x_1, x_2\}$, $x_{(2)} = \max\{x_1, x_2\}$, and $0 < y \leq 1/2$ because $0 < x_{(1)} \leq x_{(2)}$. The fitted CDF $\hat{F}(x)$ at the values $x_{(1)}$ and $x_{(2)}$ is

$$\hat{F}(x_{(1)}) = 1 - e^{-2x_{(1)}/(x_{(1)}+x_{(2)})} = 1 - e^{-2y}$$

and

$$\hat{F}(x_{(2)}) = 1 - e^{-2x_{(2)}/(x_{(1)}+x_{(2)})} = 1 - e^{-2(1-y)}.$$

It is worth noting that the fitted CDF $\hat{F}(x)$ always intersects the second riser of the empirical CDF $F_2(x)$. This is due to the fact that $\hat{F}(x_{(2)})$ can range from $1 - 1/e \cong 0.6321$ (when $y = 1/2$) to $1 - 1/e^2 \cong 0.8647$ (when $y = 0$), which are both included in the second riser's extension from 0.5 to 1. Conversely, the fitted CDF $\hat{F}(x)$ *may* intersect the first riser of the empirical CDF $F_2(x)$, depending on the value of y. When $0 < y \le \frac{\ln(2)}{2} \cong 0.3466$, the first riser is intersected by $\hat{F}(x)$ (as displayed in Figure 13.2), but when $\frac{\ln(2)}{2} < y \le 1/2$, $\hat{F}(x)$ lies entirely above the first riser (as subsequently displayed in Figure 13.6).

Define the random lengths A, B, C, and D according to the diagram in Figure 13.2. With $y = x_{(1)}/(x_{(1)} + x_{(2)})$, the lengths A, B, C and D (as functions of y) are

$$
\begin{aligned}
A &= \left(1 - e^{-2y}\right) - 0 &&= 1 - e^{-2y} && 0 < y \le 1/2,\\
B &= \left|\tfrac{1}{2} - \left(1 - e^{-2y}\right)\right| &&= \begin{cases} e^{-2y} - \frac{1}{2} & 0 < y \le \frac{\ln(2)}{2},\\ \frac{1}{2} - e^{-2y} & \frac{\ln(2)}{2} < y \le 1/2,\end{cases}\\
C &= \left(1 - e^{-2(1-y)}\right) - \tfrac{1}{2} &&= \tfrac{1}{2} - e^{-2(1-y)} && 0 < y \le 1/2,\\
D &= 1 - \left(1 - e^{-2(1-y)}\right) &&= e^{-2(1-y)} && 0 < y \le 1/2,
\end{aligned}
$$

where absolute value signs are used in the definition of B to cover the case in which $\hat{F}(x)$ does not intersect the first riser.

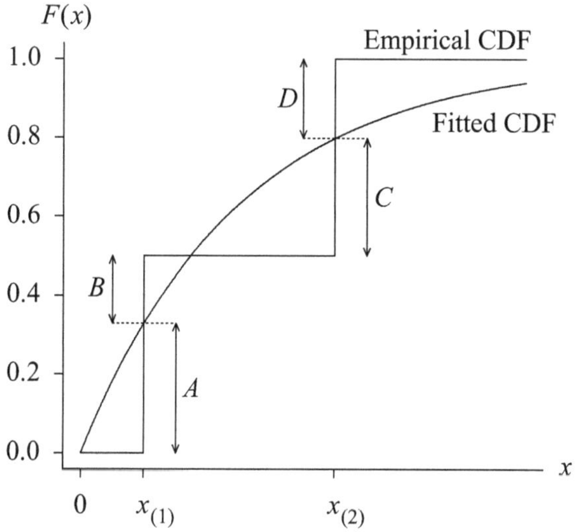

Fig. 13.2. The empirical and fitted exponential distribution for two data values $x_{(1)}$ and $x_{(2)}$. In this particular plot, $0 < y \le \frac{\ln(2)}{2}$, so the first riser of the empirical CDF $F_2(x)$ is intersected by the fitted CDF $\hat{F}(x)$

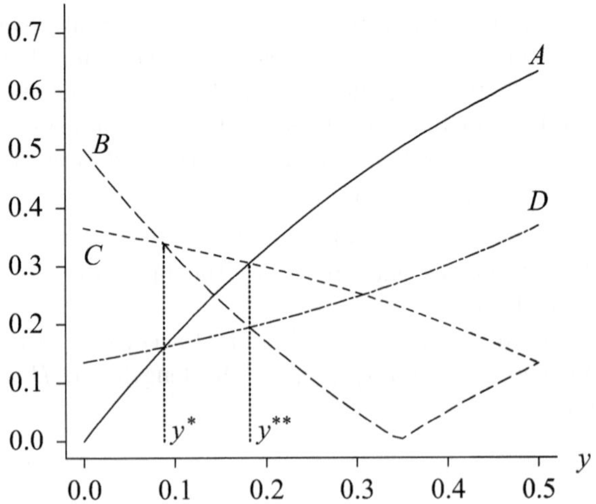

Fig. 13.3. Lengths A, B, C, and D from Figure 13.2 for $n = 2$ and $0 < y \leq 1/2$

Figure 13.3 is a graph of the lengths A, B, C, and D plotted as functions of y, for $0 < y \leq 1/2$. For any $y \in (0, 1/2]$, the K–S test statistic is $D_2 = \max\{A, B, C, D\}$. Since the length D is less than $\max\{A,\ B,\ C\}$ for all $y \in (0, 1/2]$, only A, B, C, and the y-values of their intersections (denoted by y^* and y^{**} in Figure 13.3) are needed to define D_2.

The values of y^*, $C(y^*)$, y^{**}, and $C(y^{**})$ are:

$$y^* = 1 + \frac{1}{2}\ln\left(\frac{1}{2} - \frac{1}{2}\sqrt{1 - \frac{4}{e^2}}\right) \cong 0.0880,$$

$$C(y^*) = \frac{1}{2}\sqrt{1 - \frac{4}{e^2}} \cong 0.3386,$$

$$y^{**} = 1 + \frac{1}{2}\ln\left(\frac{1}{4}\sqrt{1 + \frac{16}{e^2}} - \frac{1}{4}\right) \cong 0.1821,$$

$$C(y^{**}) = \frac{3}{4} - \frac{1}{4}\sqrt{1 + \frac{16}{e^2}} \cong 0.3052.$$

Thus, the largest vertical distance D_2 is computed using the length formula for $A(Y)$, $B(Y)$, or $C(Y)$ depending on the value of the random variable $Y = X_{(1)}/(X_{(1)} + X_{(2)})$, i.e.,

$$D_2 = \begin{cases} B(Y) & 0 < Y \leq y^* \\ C(Y) & y^* < Y \leq y^{**} \\ A(Y) & y^{**} < Y \leq 1/2. \end{cases}$$

Determining the Distribution of $\mathbf{Y} = \mathbf{X_{(1)}/(X_{(1)} + X_{(2)})}$. Let X_1, X_2 be a random sample drawn from a population having PDF

$$f(x) = \frac{1}{\theta}\, e^{-x/\theta} \qquad\qquad x > 0,$$

for $\theta > 0$. In order to determine the distribution of D_2, we must determine the distribution of $Y = X_{(1)}/(X_{(1)} + X_{(2)})$, where $X_{(1)} = \min\{X_1, X_2\}$ and $X_{(2)} = \max\{X_1, X_2\}$.

Using an order statistic result from (Hogg et al. [67, page 193]) the joint PDF of $X_{(1)}$ and $X_{(2)}$ is

$$g(x_{(1)}, x_{(2)}) = 2! \cdot \frac{1}{\theta} e^{-x_{(1)}/\theta} \cdot \frac{1}{\theta} e^{-x_{(2)}/\theta} = \left(\frac{2}{\theta^2}\right) e^{-(x_{(1)}+x_{(2)})/\theta} \quad 0 < x_{(1)} \le x_{(2)}.$$

In order to determine the PDF of $Y = X_{(1)}/(X_{(1)} + X_{(2)})$, define the dummy transformation $Z = X_{(2)}$. The random variables Y and Z define a one-to-one transformation that maps $\mathcal{A} = \{(x_{(1)}, x_{(2)}) \,|\, 0 < x_{(1)} \le x_{(2)}\}$ to $\mathcal{B} = \{(y, z) \,|\, 0 < y \le 1/2, z > 0\}$. Since $x_{(1)} = yz/(1-y)$, $x_{(2)} = z$, and the Jacobian of the inverse transformation is $z/(1-y)^2$, the joint PDF of Y and Z is

$$h(y, z) = \frac{2}{\theta^2}\, e^{-(z+yz/(1-y))/\theta} \cdot \left|\frac{z}{(1-y)^2}\right| = \frac{2z}{\theta^2(1-y)^2}\, e^{-z/(1-y)\theta}$$

for $0 < y \le 1/2$, $z > 0$. Integrating the joint PDF by parts yields the marginal PDF of Y:

$$f_Y(y) = \frac{2}{\theta^2(1-y)^2} \int_0^\infty z e^{-z/(1-y)\theta}\, dz = 2 \qquad\qquad 0 < y \le 1/2,$$

i.e., $Y \sim U(0, 1/2)$.

The final step in determining the probability distribution of D_2 is to project $\max\{A,\, B,\, C\}$ (displayed in Figure 13.4) onto the vertical axis, weighting appropriately to account for the distribution of Y. Since $\lim_{y \downarrow 0} B(y) = 1/2$, in order to determine the CDF for D_2, we must determine the functions F_α, F_β, and F_γ associated with the following intervals for the CDF of D_2:

$$F_{D_2}(d) = \begin{cases} 0 & d \le C(y^{**}) \\ F_\alpha(d) & C(y^{**}) < d \le C(y^*) \\ F_\beta(d) & C(y^*) < d \le \frac{1}{2} \\ F_\gamma(d) & \frac{1}{2} < d \le 1 - \frac{1}{e} \\ 1 & d > 1 - \frac{1}{e}. \end{cases}$$

Determining the Distribution of D_2. In order to determine F_α, F_β, and F_γ, it is necessary to find the point of intersection of a horizontal line of height $d \in [C(y^{**}), 1 - 1/e]$ with $A(y)$, $B(y)$, and $C(y)$, displayed in Figure 13.4. These points of intersection will provide integration limits for determining the distribution of D_2.

Solving each of the equations $B(y) = d$, $C(y) = d$, and $D(y) = d$ for y yields $y = -\frac{1}{2}\ln\left(d + \frac{1}{2}\right)$, $y = 1 + \frac{1}{2}\ln\left(\frac{1}{2} - d\right)$, and $y = -\frac{1}{2}\ln(1-d)$, respectively. Using these three y-values, we can determine the functions F_α, F_β, and F_γ:

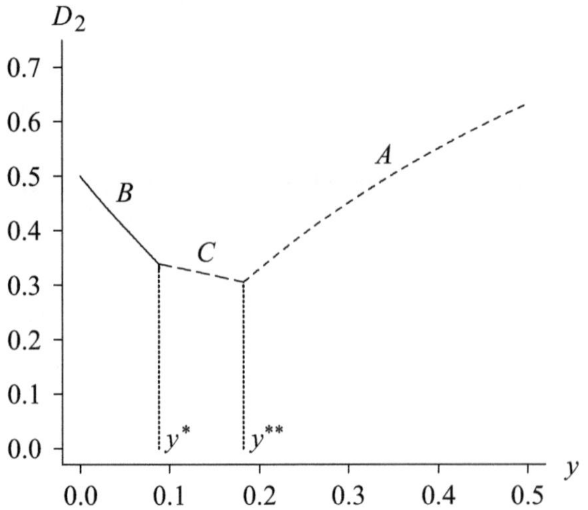

Fig. 13.4. $D_2 = \max\{A, B, C\}$ for $n = 2$ and $0 < y \le 1/2$

$$
\begin{aligned}
F_\alpha(d) &= \Pr(D_2 \le d) \\
&= \int_{1+\frac{1}{2}\ln\left(\frac{1}{2}-d\right)}^{-\frac{1}{2}\ln(1-d)} f_Y(y)\,dy \\
&= -2 - \ln[(1/2 - d)(1-d)] \qquad\qquad C(y^{**}) \le d < C(y^*), \\
F_\beta(d) &= \Pr(D_2 \le d) \\
&= \int_{-\frac{1}{2}\ln\left(d+\frac{1}{2}\right)}^{-\frac{1}{2}\ln(1-d)} f_Y(y)\,dy \\
&= \ln\left(\frac{d + 1/2}{1-d}\right) \qquad\qquad C(y^*) \le d < 1/2, \qquad \text{and} \\
F_\gamma(d) &= \Pr(D_2 \le d) \\
&= \int_0^{-\frac{1}{2}\ln(1-d)} f_Y(y)\,dy \\
&= -\ln(1-d) \qquad\qquad 1/2 \le d < 1 - 1/e.
\end{aligned}
$$

Putting the pieces together, the CDF of D_2 is

$$F_{D_2}(d) = \begin{cases} 0 & d < C(y^{**}) \\ -2 - \ln(1/2 - d) - \ln(1 - d) & C(y^{**}) \le d < C(y^*) \\ \ln(d + 1/2) - \ln(1 - d) & C(y^*) \le d < \frac{1}{2} \\ -\ln(1 - d) & \frac{1}{2} \le d < 1 - \frac{1}{e} \\ 1 & d \ge 1 - \frac{1}{e}. \end{cases}$$

Differentiating with respect to d, the PDF of D_2 is

$$f_{D_2}(d) = \begin{cases} \dfrac{1}{1-d} + \dfrac{1}{1/2+d} + \dfrac{2d}{(1/2+d)(1/2-d)} & C(y^{**}) < d < C(y^*) \\ \dfrac{1}{1-d} + \dfrac{1}{1/2+d} & C(y^*) < d < 1/2 \\ \dfrac{1}{1-d} & 1/2 < d < 1 - 1/e, \end{cases}$$

which is plotted in Figure 13.5. The percentiles of this distribution match the tabled values from Durbin [48].

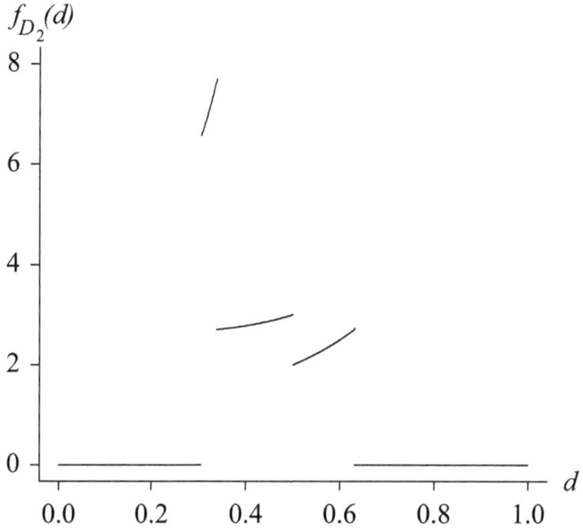

Fig. 13.5. The PDF of D_2

The distribution of D_2 can also be derived using APPL. The distribution's exact mean, variance, skewness (expected value of the standardized, centralized third moment), and kurtosis (expected value of the standardized, centralized fourth moment) can be determined in Maple with the following APPL statements:

```
> Y   := UniformRV(0, 1 / 2);
> A   := 1 - exp(-2 * y);
> B   := exp(-2 * y) - 1 / 2;
> C   := 1 / 2 - exp(-2 * (1 - y));
> ys  := solve(B = C, y)[1];
> yss := solve(A = C, y)[1];
> g   := [[unapply(B, y), unapply(C, y), unapply(A, y)],
>          [0, ys, yss, 1 / 2]];
> D2  := Transform(Y, g);
> Mean(D2);
> Variance(D2);
> Skewness(D2);
> Kurtosis(D2);
```

The Maple `solve` procedure is used to find y^* and y^{**} (the variables `ys` and `yss`) and the APPL `Transform` procedure transforms the random variable Y to D_2 using the piecewise segments B, C, and A. The expressions for the mean, variance, skewness, and kurtosis are given in terms of radicals, exponentials, and logarithms, e.g.,

$$
\begin{aligned}
E[D_2] &= -1 + \left(\frac{1}{2e}\right) 2 - r - 2s \\
&\quad + \left(\frac{1}{2}\right) 2\ln(e^2 + er) + 6\ln 2 + \ln(er - e^2) - \ln(e^2 - es) + \ln(e^2 + es),
\end{aligned}
$$

where $r = \sqrt{e^2 + 16}$ and $s = \sqrt{e^2 - 4}$. The others are too lengthy to display here, but the decimal approximations for the mean, variance, skewness, and kurtosis are, respectively, $E(D_2) \cong 0.4430$, $V(D_2) \cong 0.0100$, $\gamma_3 \cong 0.2877$, and $\gamma_4 \cong 1.7907$.

Example 13.1. Suppose that two data values, $x_{(1)} = 95$ and $x_{(2)} = 100$, constitute a random sample from an unknown population. The hypothesis test

$$
\begin{aligned}
H_0 &: F(x) = F_0(x) \\
H_1 &: F(x) \neq F_0(x),
\end{aligned}
$$

where $F_0(x) = 1 - e^{-x/\theta}$, is used to test the legitimacy of modeling the data set with an exponential distribution. The MLE is $\hat{\theta} = (95 + 100)/2 = 97.5$. The empirical distribution function, fitted exponential distribution, and corresponding lengths A, B, C, and D are displayed in Figure 13.6.
The ratio $y = x_{(1)}/(x_{(1)} + x_{(2)}) = 95/195$ corresponds to A being the maximum of A, B, C, and D. This yields the test statistic

$$
d_2 = 1 - e^{-2(95/195)} \cong 0.6226,
$$

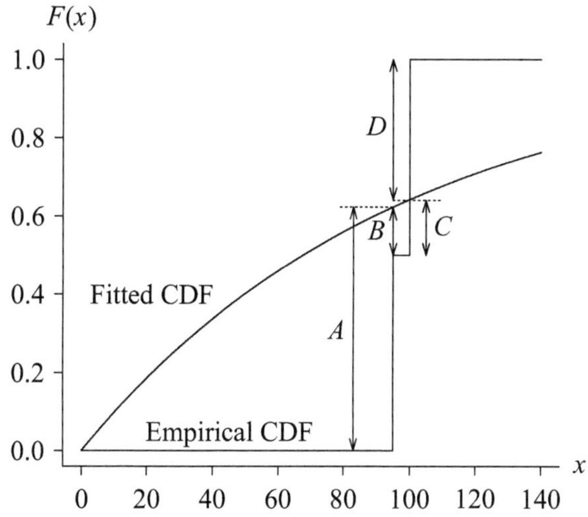

Fig. 13.6. The empirical and fitted exponential distribution for two data values $x_{(1)} = 95$ and $x_{(2)} = 100$. In this example, $y > \frac{\ln(2)}{2}$, so the first riser of the empirical CDF is not intersected by the fitted CDF

which falls in the right-hand tail of the distribution of D_2, as displayed in Figure 13.7. (The two breakpoints in the CDF are also indicated in Figure 13.7.) Hence, the test statistic provides evidence to reject the null hypothesis for the goodness-of-fit test. Since large values of the test statistic lead to rejecting H_0, the p-value associated with this particular data set is

$$p = 1 - F_{D_2}\left(1 - e^{-2(95/195)}\right) = 1 + \ln\left(1 - \left[1 - e^{-2(95/195)}\right]\right) = \frac{1}{39} \cong 0.02564.$$

Using the exact PDF of D_2 to determine the p-value is superior to using tables (e.g., Durbin [48]) since the approximation associated with linear interpolation is avoided. The exact PDF is also preferred to approximations from Stephens [152] and Law [88], which often do not perform well for small values of n. Since the distribution of D_2 is not a function of θ, the power of the hypothesis test as a function of θ is constant with a value of $1 - \alpha$. Since there appears to be a pattern to the functional forms associated with the three segments of the PDF of D_2, we derive the distribution of D_3 in the appendix in an attempt to establish a pattern.

The focus of the paper now shifts to investigating the distributions of other goodness-of-fit statistics.

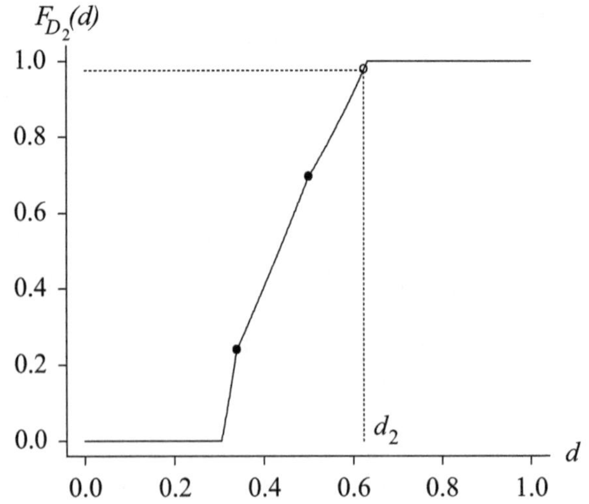

Fig. 13.7. The CDF of D_2 and the test statistic $D_2 = 1 - e^{-2(95/195)} \cong 0.6226$

13.2 Other Measures of Fit

The K–S test statistic measures the distance between $F_n(x)$ and $\hat{F}(x)$ by using the L_∞ norm. The square of the L_2 norm gives the test statistic

$$L_2^2 = \int_{-\infty}^{\infty} \left(F_n(x) - \hat{F}(x)\right)^2 dx,$$

which, for exponential sampling and $n = 1$ data value, is

$$L_2^2 = \int_0^{x_1} \left(1 - e^{-x/x_1}\right)^2 dx + \int_{x_1}^{\infty} e^{-2x/x_1}\, dx = \left(\frac{4-e}{2e}\right) x_1.$$

Since $X_1 \sim \text{exponential}(\theta)$, $L_2^2 \sim \text{exponential}\left(\frac{4-e}{2e}\,\theta\right)$. Unlike the K–S test statistic, the square of the L_2 norm is dependent on θ. For exponential sampling with $n = 2$, the square of the L_2 norm is also dependent on θ:

$$\begin{aligned}
L_2^2 &= \int_0^{\infty} \left(F_2(x) - \hat{F}(x)\right)^2 dx \\
&= \int_0^{x_{(1)}} \hat{F}(x)^2\, dx + \int_{x_{(1)}}^{x_{(2)}} \left(\hat{F}(x) - \frac{1}{2}\right)^2 dx + \int_{x_{(2)}}^{\infty} \left(1 - \hat{F}(x)\right)^2 dx \\
&= \int_0^{x_{(1)}} \left(1 - e^{-2x/(x_1+x_2)}\right)^2 dx + \int_{x_{(1)}}^{x_{(2)}} \left(\frac{1}{2} - e^{-2x/(x_1+x_2)}\right)^2 dx \\
&\quad + \int_{x_{(2)}}^{\infty} \left(e^{-2x/(x_1+x_2)}\right)^2 dx \\
&= -\frac{x_{(2)}}{2} + \frac{x_{(1)} + x_{(2)}}{2} \cdot \left(e^{-2x_{(1)}/\left(x_{(1)}+x_{(2)}\right)} + e^{-2x_{(2)}/\left(x_{(1)}+x_{(2)}\right)}\right),
\end{aligned}$$

where $X_{(1)} \sim$ exponential(2θ) and $X_{(2)}$ has the marginal PDF

$$f_{X_{(2)}}(x) = 2\left(1 - e^{-x/\theta}\right)\left(\frac{1}{\theta}e^{-x/\theta}\right), \qquad x > 0.$$

Unlike the square of the L_2 norm, the Cramer–von Mises and Anderson–Darling test statistics (see Lawless [91]) are distribution-free. They can be defined as

$$W_n^2 = n\int_{-\infty}^{\infty}\left(F_n(x) - \hat{F}(x)\right)^2 d\hat{F}(x)$$

and

$$A_n^2 = n\int_{-\infty}^{\infty}\frac{\left(F_n(x) - \hat{F}(x)\right)^2}{\hat{F}(x)\left[1 - \hat{F}(x)\right]}\, d\hat{F}(x),$$

where n is the sample size. The computational formulas for these statistics are

$$W_n^2 = \sum_{i=1}^{n}\left(\hat{F}(x_{(i)}) - \frac{i - 0.5}{n}\right)^2 + \frac{1}{12n}$$

and

$$A_n^2 = -\sum_{i=1}^{n}\frac{2i-1}{n}\left(\ln\left(\hat{F}(x_{(i)})\right) + \ln\left(1 - \hat{F}(x_{(n+1-i)})\right)\right) - n.$$

13.2.1 Distribution of W_1^2 and A_1^2 for Exponential Sampling

When $n = 1$ and sampling is from an exponential population, the Cramer–von Mises test statistic is

$$W_1^2 = \left(\frac{1}{2} - \frac{1}{e}\right)^2 + \frac{1}{12} = \frac{1}{3} - \frac{1}{e} + \frac{1}{e^2}.$$

Thus, the Cramer–von Mises test statistic is degenerate for $n = 1$ with CDF

$$F_{W_1^2}(w) = \begin{cases} 0 & w < \dfrac{1}{3} - \dfrac{1}{e} + \dfrac{1}{e^2} \\ 1 & w \ge \dfrac{1}{3} - \dfrac{1}{e} + \dfrac{1}{e^2}. \end{cases}$$

When $n = 1$ and sampling is from an exponential population, the Anderson–Darling test statistic is

$$A_1^2 = -\ln(1 - e^{-1}) - \ln(e^{-1}) - 1 = 1 - \ln(e - 1).$$

It is also degenerate for $n = 1$ with CDF

$$F_{A_1^2}(a) = \begin{cases} 0 & a < 1 - \ln(e - 1) \\ 1 & a \ge 1 - \ln(e - 1). \end{cases}$$

13.2.2 Distribution of W_2^2 and A_2^2 for Exponential Sampling

When $n = 2$ and sampling is from an exponential population, the Cramer–von Mises test statistic is

$$W_2^2 = \left(e^{-x_{(1)}/\hat{\theta}} - \frac{3}{4}\right)^2 + \left(e^{-x_{(2)}/\hat{\theta}} - \frac{1}{4}\right)^2 + \frac{1}{24},$$

where $\hat{\theta} = (x_1 + x_2)/2$. The Anderson–Darling test statistic is

$$A_2^2 = 2 - \frac{1}{2}\ln\left(e^{x_{(1)}/\hat{\theta}} - 1\right) - \frac{3}{2}\ln\left(e^{x_{(2)}/\hat{\theta}} - 1\right).$$

If we let $y = x_{(1)}/(x_{(1)} + x_{(2)})$, as we did when working with D_2, we obtain the following formulas for W_2^2 and A_2^2 in terms of y,

$$W_2^2 = \left(e^{-2y} - \frac{3}{4}\right)^2 + \left(e^{-2(1-y)} - \frac{1}{4}\right)^2 + \frac{1}{24},$$

and

$$A_2^2 = 2 - \frac{1}{2}\ln\left(e^{2y} - 1\right) - \frac{3}{2}\ln\left(e^{2(1-y)} - 1\right),$$

for $0 < y \le 1/2$. Graphs of D_2, W_2^2, and A_2^2 are displayed in Figure 13.8. Although the ranges of the three functions are quite different, they share similar shapes.

Each of the three functions plotted in Figure 13.8 achieves a minimum between $y = 0.15$ and $y = 0.2$. The Cramer–von Mises test statistic W_2^2 achieves a minimum at y^{***} that satisfies

$$4e^{-4y} - 3e^{-2y} - 4e^{-4(1-y)} + e^{-2(1-y)} = 0$$

for $0 < y \le 1/2$. This is equivalent to fourth-degree polynomial in e^{2y} that can be solved exactly using radicals. The minimum is achieved at $y^{***} \cong 0.1549$.

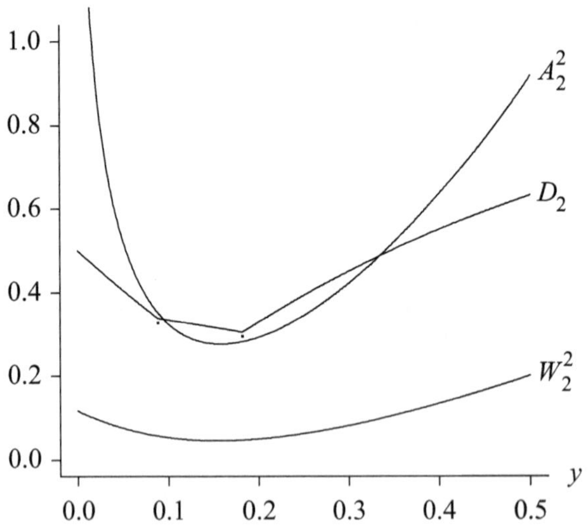

Fig. 13.8. Graphs of D_2, W_2^2, and A_2^2 for $n = 2$ and $0 < y \le 1/2$

Likewise, the Anderson–Darling test statistic A_2^2 achieves a minimum at y^{****} that satisfies

$$e^{2y} + 2e^2 - 3e^{2(1-y)} = 0$$

for $0 < y \le 1/2$, which yields

$$y^{****} = \frac{1}{2} + \frac{1}{2}\ln\left(\sqrt{e^2+3} - e\right) \cong 0.1583.$$

These values and other pertinent values associated with D_2, W_2^2, and A_2^2 are summarized in Table 13.1.

Table 13.1. Pertinent values associated with the test statistics D_2, W_2^2, and A_2^2

Test statistic	Value when $y = 0$	Minimized at	Global minimum on $(0, 1/2]$	Value when $y = 1/2$
D_2	$\frac{1}{2}$	$y^{**} \cong 0.1821$	$D_2(y^{**}) \cong 0.3052$	$1 - \frac{1}{e} \cong 0.6321$
W_2^2	$\frac{1}{6} + \frac{1}{e^4} - \frac{1}{2e^2}$ $\cong 0.1173$	$y^{***} \cong 0.1549$	$W_2^2(y^{***}) \cong 0.04623$	$\frac{2}{e^2} - \frac{2}{e} + \frac{2}{3} \cong 0.2016$
A_2^2	$+\infty$	$y^{****} \cong 0.1583$	$A_2^2(y^{****}) \cong 0.2769$	$2 - 2\ln(e-1) \cong 0.9174$

Figure 13.8 can be helpful in determining which of the three goodness-of-fit statistics is appropriate in a particular application. Consider, for instance, a reliability engineer who is interested in detecting whether reliability growth or reliability degradation is occurring for a repairable system. One would expect a shorter failure time followed by a longer failure time if reliability growth were occurring; one would expect a longer failure time followed by a shorter failure time if reliability degradation were occurring. In either case, these correspond to a small value of y, so Figure 13.8 indicates that the Anderson–Darling test is preferred due to the vertical asymptote at $y = 0$.

For notational convenience below, let $D_2(y)$, $W_2^2(y)$, and $A_2^2(y)$ denote the values of D_2, W_2^2, and A_2^2, respectively, corresponding to a specific value of y. For example, $W_2^2(1/4)$ denotes the value of W_2^2 when $y = 1/4$.

Example 13.2. Consider again the data from Example 13.1: $x_{(1)} = 95$ and $x_{(2)} = 100$. Since $y = 95/195 = 19/39$, the Cramer–von Mises test statistic is

$$w_2^2 = \left(e^{-38/39} - 3/4\right)^2 + \left(e^{-40/39} - 1/4\right)^2 + 1/24 \cong 0.1923.$$

The p-value for this test statistic is the same as the p-value for the K–S test statistic, namely

$$\int_{19/39}^{1/2} f_Y(y)dy = \int_{19/39}^{1/2} 2dy = 1/39 \cong 0.0256.$$

More generally, for each value of y such that both $D_2(y) > D_2(0)$ and $W_2^2(y) > W_2^2(0)$,

$$F_{D_2}\left(D_2(y)\right) = F_Y\left(y\right) = F_{W_2^2}\left(W_2^2(y)\right).$$

The Anderson–Darling test statistic for $x_{(1)} = 95$ and $x_{(2)} = 100$ is

$$a_2^2 = 2 - \frac{1}{2}\ln\left(e^{38/39} - 1\right) - \frac{3}{2}\ln\left(e^{40/39} - 1\right) \cong 0.8774.$$

Since the value of $A_2^2(y)$ exceeds the test statistic $a_2^2 \cong 0.8774$ only for $y < y' = 0.02044$ [where y' is the first intersection point of $A_2^2(y)$ and the horizontal line with height $A_2^2(19/39)$] and for $y > 19/39$, the p-value for the Anderson–Darling goodness-of-fit test is given by

$$p = \int_0^{y'} 2\,dy + \int_{19/39}^{1/2} 2\,dy = 2y' + (1 - 38/39) \cong 0.06654.$$

Determining the Distribution of W_2^2 and A_2^2. As was the case with D_2, we can find exact expressions for the PDFs of W_2^2 and A_2^2. Consider W_2^2 first. For w values in the interval $W_2^2(y^{***}) \le w < W_2^2(0)$, the CDF of W_2^2 is

$$F_{W_2^2}(w) = \int_{y_1}^{y_2} f_Y(y)\,dy = 2(y_2 - y_1),$$

where y_1 and y_2 are the ordered solutions to $W_2^2 = w$. For w values in the interval $W_2^2(0) \le w < W_2^2(1/2)$, the CDF of W_2^2 is

$$F_{W_2^2}(w) = \int_0^{y_1} f_Y(y)\,dy = 2y_1,$$

where y_1 is the solution to $W_2^2 = w$ on $0 < y < 1/2$. The following APPL code can be used to find the PDF of W_2^2.

```
> Y    := UniformRV(0, 1 / 2);
> W    := (exp(-2 * y) - 3 / 4) ^ 2
>           + (exp(-2 * (1 - y)) - 1 / 4) ^ 2 + 1 / 24;
> ysss := solve(diff(W, y) = 0, y)[1];
> g    := [[unapply(W, y), unapply(W, y)], [0, ysss, 1 / 2]];
> W2   := Transform(Y, g);
```

The `Transform` procedure requires that the transformation `g` be input in piecewise monotone segments. The resulting PDF for W_2^2 is too lengthy to display here.

Now consider A_2^2. For a value in the interval $A_2^2(y^{****}) \le a < A_2^2(1/2)$, the CDF of A_2^2 is

$$F_{A_2^2}(a) = \int_{y_1}^{y_2} f_Y(y)\,dy = 2(y_2 - y_1),$$

where y_1 and y_2 are the ordered solutions to $A_2^2 = a$ on $0 < y \le 1/2$. For a values in the interval $A_2^2(1/2) \le a < \infty$, the CDF of A_2^2 is

$$F_{A_2^2}(a) = \int_{y_1}^{1/2} f_Y(y)\,dy = 1 - 2y_1,$$

where y_1 is the solution to $A_2^2 = a$ on $0 < y < 1/2$. The following APPL code can be used to find the PDF of A_2^2.

```
> Y     := UniformRV(0, 1 / 2);
> A     := 2 - ln(exp(2 * y) - 1) / 2 - 3
>            * ln(exp(2 * (1 - y)) - 1) / 2;
> yssss := solve(diff(A, y) = 0, y)[1];
> g     := [[unapply(A, y), unapply(A, y)], [0, yssss, 1/2]];
> A2    := Transform(Y, g);
```

The resulting PDF for A_2^2 is again too lengthy to display here.

13.3 Applications

Although statisticians prefer large sample sizes because of the associated desirable statistical properties of estimators as the sample size n becomes large, there are examples of real-world data sets with only $n = 2$ observations in which the fit to an exponential distribution is important. In this section, we focus on two applications: U.S. Space Shuttle flights and the world-wide commercial nuclear power industry. Both applications involve significant government expenditures associated with decisions that must be made based on limited data. In both cases, "events" are failures and the desire is to test whether a homogeneous Poisson process model or other (e.g., a nonhomogeneous Poisson process) model is appropriate, i.e., determining whether failures occur randomly over time. Deciding which of the models is appropriate is important to reliability engineers since a nonhomogeneous Poisson process with a decreasing intensity function may be a sign of reliability growth or improvement over time (Rigdon and Basu [135]).

Example 13.3. NASA's Space Shuttle program has experienced $n = 2$ catastrophic failures which have implications for the way in which the United States will pursue future space exploration. On January 28, 1986, the *Challenger* exploded 72 s after liftoff. Failure of an O-ring was determined as the most likely cause of the accident. On February 1, 2003, Shuttle *Columbia* was lost during its return to Earth. Investigators believed that tile damage during ascent caused the accident. These two failures occurred on the 25th and 113th Shuttle flights. A goodness-of-fit test is appropriate to determine whether the failures occurred randomly, or equivalently, whether a Poisson process model is appropriate. The hope is that the data will *fail* this test due to the fact that reliability growth has occurred due to the many improvements that have been made to the Shuttle (particularly after the *Challenger* accident), and perhaps a nonhomogeneous Poisson process with a decreasing intensity function is a more appropriate stochastic model for failure times. Certainly, large amounts of money have been spent and some judgments about the safety and direction of the future of the Shuttle program should be made on the basis of these two data values.

The appropriate manner to model *time* in this application is nontrivial. There is almost certainly increased risk on liftoff and landing, but the time spent on the mission should also be included because an increased mission time means an increased exposure to internal and external failures while a Shuttle is in orbit. Because of this inherent difficulty in quantifying time, we do our numerical analysis on an example in an application area where time is more easily measured.

Example 13.4. The world-wide commercial nuclear power industry has experienced $n = 2$ core meltdowns in its history. The first was at the Three Mile Island nuclear facility on March 28, 1979. The second was at Chernobyl on April 26, 1986. As in the case of the Space Shuttle accidents, it is again of interest to know whether the meltdowns can be considered to be events from a Poisson process. The hypothesis test of interest here is whether the two times to meltdown are independent observations from an exponential population with a rate parameter estimated from data. Measuring time in this case is not trivial because of the commissioning and decommissioning of facilities over time. The first nuclear power plant was the Calder Hall I facility in the United Kingdom, commissioned on October 1, 1956. Figure 13.9 shows the evolution of the number of active commercial reactors between that date and the Chernobyl accident on April 26, 1986. The commissioning and decommissioning dates of all commercial nuclear reactors is given in Cho and Spiegelberg–Planer [32]. Downtime for maintenance has been ignored in determining the times of the two accidents.

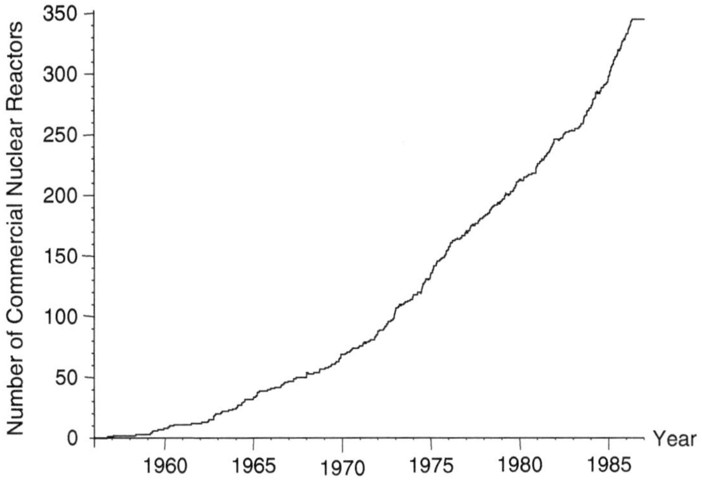

Fig. 13.9. Number of operating commercial nuclear power plants world-wide between October 1, 1956 and April 26, 1986

Using the data illustrated in Figure 13.9, the time of the two accidents measured in cumulative commercial nuclear reactor years is found by integrating under the curve. The calendar dates were converted to decimal values using Julian dates, adjusting for leap years. The two accidents occurred at 1548.02 and 3372.27 cumulative operating years, respectively. This means that the hypothesis test is to see whether the times between accidents, namely 1548.02 and $3372.27 - 1548.02 = 1824.25$ years can be considered independent observations from an exponential population. The maximum likelihood estimator for the mean time between core meltdowns is $\hat{\theta} = 1686.14$ years. This results in a y-value of $y = 1548.02/(1548.02 + 1824.25) = 0.459$ and a K–S test statistic of $d_2 = 0.601$. This corresponds to a p-value for the associated goodness-of-fit test of $p = 1 + \ln(1 - d_2) = 0.082$. There is *not* enough statistical evidence to conclude a nonhomogeneous model is appropriate here, so it is reasonable to model nuclear power accidents as random events. Figure 13.10 shows a plot of the fitted CDF and associated values of A, B, C, and D for this data set.

Acknowledgments

The first author acknowledges summer support from Rose–Hulman Institute of Technology. The second and third authors acknowledge FRA support from the College of William & Mary. The authors also acknowledge the assistance of Bill Griffith, Thom Huber, and David Kelton in selecting data sets for the case studies in Section 13.3.

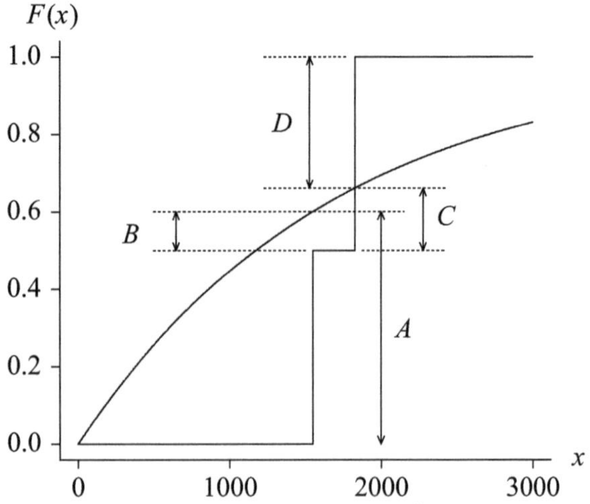

Fig. 13.10. The empirical and fitted exponential distribution for the two times between nuclear reactor meltdowns (in years), $x_{(1)} = 1548.02$ and $x_{(2)} = 1824.25$

Appendix: Distribution of D_3 for Exponential Sampling

The pattern that emerged in the piecewise representation of the PDF of D_2 led us to derive the PDF of D_3 to see if any similar patterns arose. This appendix contains a derivation of the distribution of the K–S test statistic when $n = 3$ observations x_1, x_2, and x_3 are drawn from an exponential population with fixed, positive, unknown mean θ. The maximum likelihood estimator is $\hat{\theta} = (x_1 + x_2 + x_3)/3$, which results in the fitted CDF

$$\hat{F}(x) = 1 - e^{-x/\hat{\theta}} \qquad x > 0.$$

Analogous to the $n = 2$ case, define

$$y = \frac{x_{(1)}}{x_{(1)} + x_{(2)} + x_{(3)}}$$

and

$$z = \frac{x_{(2)}}{x_{(1)} + x_{(2)} + x_{(3)}}$$

so that

$$1 - y - z = \frac{x_{(3)}}{x_{(1)} + x_{(2)} + x_{(3)}}.$$

The domain of definition of y and z is

$$\mathcal{D} = \{(y, z) \,|\, 0 < y < z < (1 - y)/2\}.$$

The values of the fitted CDF at the three order statistics are

$$\hat{F}(x_{(1)}) = 1 - e^{-x_{(1)}/\hat{\theta}} = 1 - e^{-3y},$$

$$\hat{F}(x_{(2)}) = 1 - e^{-x_{(2)}/\hat{\theta}} = 1 - e^{-3z},$$

and

$$\hat{F}(x_{(3)}) = 1 - e^{-x_{(3)}/\hat{\theta}} = 1 - e^{-3(1-y-z)}.$$

The vertical distances A, B, C, D, E, and F (as functions of y and z) are defined in a similar fashion to the $n = 2$ case (see Figure 13.2):

$$\begin{aligned}
A &= 1 - e^{-3y} \\
B &= \left|\tfrac{1}{3} - \left(1 - e^{-3y}\right)\right| &&= \left|e^{-3y} - \tfrac{2}{3}\right| \\
C &= \left|\left(1 - e^{-3z}\right) - \tfrac{1}{3}\right| &&= \left|e^{-3z} - \tfrac{2}{3}\right| \\
D &= \left|\tfrac{2}{3} - \left(1 - e^{-3z}\right)\right| &&= \left|e^{-3z} - \tfrac{1}{3}\right| \\
E &= \left|\left(1 - e^{-3(1-y-z)}\right) - \tfrac{2}{3}\right| &&= \left|e^{-3(1-y-z)} - \tfrac{1}{3}\right| \\
F &= 1 - \left(1 - e^{-3(1-y-z)}\right) &&= e^{-3(1-y-z)}
\end{aligned}$$

for $(y, z) \in \mathcal{D}$.

Figure 13.11 shows the regions associated with the maximum of A, B, C, D, E, F for $(y, z) \in \mathcal{D}$. In three dimensions, with $D_3 = \max\{A, B, C, D, E, F\}$ as the third axis, this figure appears to be a container with the region E at the bottom of the container and with each of the other four sides rising as they move away from their intersection with E. The absolute value signs that appear in the final formulas for B, C, D, and E above can be easily removed since, over the region $\mathcal{D}$ associated with D_3, the expressions within the absolute value signs are always positive for B and D, but always negative for C and E. The distance F is never the largest of the six distances for any $(y, z) \in \mathcal{D}$, so it can be excluded from consideration. Table 13.2 gives the functional forms of the two-way intersections between the five regions shown in Figure 13.11. Note that the BC and AD curves, and the AC and BD curves, are identical.

In order to determine the breakpoints in the support for D_3, it is necessary to find the (y, z) coordinates of the three-way intersections of the five regions in Figure 13.11 and the two-way intersections of the regions on the boundary of $\mathcal{D}$. Table 13.3 gives the values of y and z for these breakpoints on the boundary of $\mathcal{D}$, along with the value of $D_3 = \max\{A, B, C, D, E, F\}$ at these values, beginning at $(y, z) = (0, 1/2)$ and proceeding in a counterclockwise direction. One point has been excluded from Table 13.3 because of the intractability of the values (y, z). The three-way intersection between regions A, C, and the line $z = (1 - y)/2$ can only be expressed in terms of the solution to a cubic equation. After some algebra, the point of intersection is the decimal approximation $(y, z) \cong (0.1608, 0.4196)$ and the associated value of D_3 is $2/3$ minus the only real solution to the cubic equation

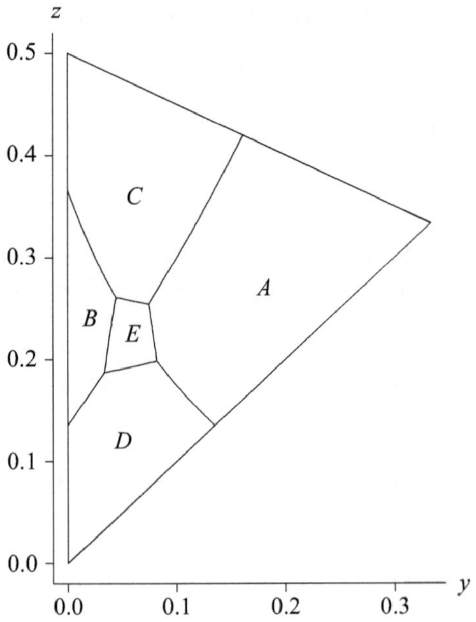

Fig. 13.11. Regions associated with $\max\{A, B, C, D, E, F\}$ over $(y, z) \in \mathcal{D}$

Table 13.2. Intersections of regions A, B, C, D, and E in $\mathcal{D}$

Region intersection	Function
AD	$z = -\frac{1}{3}\ln\left(\frac{4}{3} - e^{-3y}\right)$
BD	$z = -\frac{1}{3}\ln\left(e^{-3y} - \frac{1}{3}\right)$
BC	$z = -\frac{1}{3}\ln\left(\frac{4}{3} - e^{-3y}\right)$
AC	$z = -\frac{1}{3}\ln\left(e^{-3y} - \frac{1}{3}\right)$
AE	$z = \frac{1}{3}\ln\left[e^{3(1-y)}\left(e^{-3y} - \frac{2}{3}\right)\right]$
DE	$z = \frac{1}{3}\ln\left[\frac{1}{3}e^{3(1-y)}\left(1 - \sqrt{1 - 9e^{-3(1-y)}}\right)\right]$
BE	$z = \frac{1}{3}\ln\left[e^{3(1-y)}\left(1 - e^{-3y}\right)\right]$
CE	$z = \frac{1}{3}\ln\left[\frac{1}{6}e^{3(1-y)}\left(-1 + \sqrt{1 + 36e^{-3(1-y)}}\right)\right]$

$$3d^3 + d^2 - 3e^{-3} = 0,$$

which yields

$$d_{AC} = \frac{7}{9} - \frac{1}{18}\left(2916e^{-3} - 8 + c\right)^{1/3} - \frac{2}{9}\left(2916e^{-3} - 8 + c\right)^{-1/3} \cong 0.3827,$$

where $c = 108\sqrt{729e^{-6} - 4e^{-3}}$.

Table 13.3. Intersection points along the boundary of $\mathcal{D}$

y	z	D_3
0	1/2	$2/3 - e^{-3/2} \cong 0.4435$
0	ln(3)/3	$1/3 \cong 0.3333$
0	ln(3/2)/3	$1/3 \cong 0.3333$
0	0	$2/3 \cong 0.6667$
ln(3/2)/3	ln(3/2)/3	$1/3 \cong 0.3333$
1/3	1/2	$1 - 1/e \cong 0.6321$

The three-way intersection points in the interior of $\mathcal{D}$ are more difficult to determine than those on the boundary. The value of D_3 associated with each of these four points is the single real root of a cubic equation on the support of D_3. These equations and approximate solution values, in ascending order, are given in Table 13.4. For example, consider the value of the maximum at the intersection of regions A, C, and E in Figure 13.11. The value of D_3 must satisfy the cubic equation

$$e^3 (1-d)\left(\frac{2}{3} - d\right)\left(\frac{1}{3} - d\right) = 1,$$

which yields

$$d_{ACE} = \frac{(243+c)^{2/3}\, 12^{2/3} c - 243\,(243+c)^{2/3}\, 12^{2/3} + 144e^5 - 12^{4/3} e^4 (243+c)^{1/3}}{216e^5},$$

or approximately $d_{ACE} \cong 0.19998$, in which $c = \sqrt{59049 - 12e^6}$.

Table 13.4. Three-way interior intersection points of regions A, B, C, D, and E in $\mathcal{D}$

Regions	Cubic equation	Approximate solution
ACE	$e^3 (1-d)\left(\frac{2}{3}-d\right)\left(\frac{1}{3}-d\right) = 1$	$d_{ACE} \cong 0.2000$
BCE	$e^3 \left(\frac{1}{3}-d\right)\left(d+\frac{2}{3}\right)\left(\frac{2}{3}-d\right) = 1$	$d_{BCE} \cong 0.2091$
ADE	$e^3 \left(\frac{1}{3}-d\right)\left(d+\frac{1}{3}\right)(1-d) = 1$	$d_{ADE} \cong 0.2178$
BDE	$e^3 \left(d+\frac{2}{3}\right)\left(d+\frac{1}{3}\right)\left(\frac{1}{3}-d\right) = 1$	$d_{BDE} \cong 0.2366$

The largest value of $D_3 = \max\{A, B, C, D, E\}$ on $\mathcal{D}$ occurs at the origin ($y = 0$ and $z = 0$) and has value 2/3, which is the upper limit of the support of D_3. The smallest value of D_3 on $\mathcal{D}$ occurs at the intersection ACE and is $d_{ACE} \cong 0.19998$, which is the lower limit of the support of D_3.

Determining the Joint Distribution of Y and Z. The next step is to determine the distribution of $Y = X_{(1)}/(X_{(1)} + X_{(2)} + X_{(3)})$ and $Z = X_{(2)}/(X_{(1)} + X_{(2)} + X_{(3)})$. Using an order statistic result from Hogg et al. [67, page 193], the joint PDF of $X_{(1)}$, $X_{(2)}$, and $X_{(3)}$ is

$$g(x_{(1)}, x_{(2)}, x_{(3)}) = \frac{3!}{\theta^3} \exp(-(x_{(1)} + x_{(2)} + x_{(3)})/\theta) \qquad 0 < x_{(1)} \le x_{(2)} \le x_{(3)}.$$

In order to determine the joint PDF of $Y = X_{(1)}/(X_{(1)} + X_{(2)} + X_{(3)})$ and $Z = X_{(2)}/(X_{(1)} + X_{(2)} + X_{(3)})$, define the dummy transformation $W = X_{(3)}$. The random variables Y, Z, and W define a one-to-one transformation from $\mathcal{A} = \{(x_{(1)}, x_{(2)}, x_{(3)}) \,|\, 0 < x_{(1)} \le x_{(2)} \le x_{(3)})\}$ to $\mathcal{B} = \{(y, z, w) \,|\, 0 < y < z < (1 - y)/2, w > 0\}$. Since $x_{(1)} = yw/(1 - y - z)$, $x_{(2)} = zw/(1 - y - z)$, and $x_{(3)} = w$, and the Jacobian of the inverse transformation is $w^2/(1 - y - z)^3$, the joint PDF of Y, Z, and W on $\mathcal{B}$ is

$$\begin{aligned} h(y, z, w) &= \frac{6}{\theta^3} \exp\left(-\left(\frac{yw + zw}{1 - y - z} + w\right)/\theta\right) \left|\frac{w^2}{(1 - y - z)^3}\right| \\ &= \frac{6w^2}{\theta^3 (1 - y - z)^3} \exp\left(-\frac{w}{(1 - y - z)\theta}\right) \qquad (y, z, w) \in \mathcal{B}. \end{aligned}$$

Integrating by parts, the joint PDF of Y and Z on $\mathcal{D}$ is

$$f_{Y,Z}(y, z) = \frac{6}{\theta^3 (1 - y - z)^3} \int_0^\infty w^2 \exp\left(-\frac{w}{(1 - y - z)\theta}\right) dw = 12 \quad (y, z, w) \in \mathcal{D},$$

i.e., Y and Z are uniformly distributed on $\mathcal{D}$.

Determining the Distribution of D_3. The CDF of D_3 will be defined in a piecewise manner, with breakpoints at the following ordered quantities: d_{ACE}, d_{BCE}, d_{ADE}, d_{BDE}, $1/3$, d_{AC}, $\frac{2}{3} - e^{-3/2}$, $1 - \frac{1}{e}$, and $2/3$. The CDF $F_{D_3}(d) = \Pr(D_3 \le d)$ is found by integrating the joint PDF of Y and Z over the appropriate limits, yielding

$$F_{D_3}(d) = \begin{cases} 0 & d < d_{ACE} \\ \frac{2}{3}\left[\ln\left(e^3[1-d]\left[\frac{2}{3}-d\right]\left[\frac{1}{3}-d\right]\right)\right]^2 & d_{ACE} \le d < d_{BCE} \\ \frac{2}{3}\ln\left[e^6(1-d)\left(\frac{2}{3}-d\right)^2\left(\frac{2}{3}+d\right)\left(\frac{1}{3}-d\right)^2\right] & \\ \qquad \times \ln\left(\frac{1-d}{2/3+d}\right) & d_{BCE} \le d < d_{ADE} \\ \frac{4}{3}\ln\left(\frac{d+1/3}{2/3-d}\right)\ln\left(\frac{d+2/3}{1-d}\right) & \\ \qquad -\frac{2}{3}\left[\ln\left(e^3\left[d+\frac{2}{3}\right]\left[d+\frac{1}{3}\right]\left[\frac{1}{3}-d\right]\right)\right]^2 & d_{ADE} \le d < d_{BDE} \\ \frac{4}{3}\ln\left(\frac{d+1/3}{2/3-d}\right)\ln\left(\frac{d+2/3}{1-d}\right) & d_{BDE} \le d < \frac{1}{3} \\ \frac{4}{3}\ln\left(\frac{2/3-d}{d+1/3}\right)\ln(1-d) - \frac{2}{3}\left[\ln\left(\frac{d+1/3}{1-d}\right)\right]^2 & \frac{1}{3} \le d < d_{AC} \\ 1-\frac{2}{3}\left[\ln\left(d+\frac{1}{3}\right)\right]^2 - \left[1+\ln(1-d)\right]^2 & \\ \qquad -3\left[1+\frac{2}{3}\ln\left(\frac{2}{3}-d\right)\right]^2 & d_{AC} \le d < \frac{2}{3} - e^{-3/2} \\ 1-\frac{2}{3}\left[\ln\left(d+\frac{1}{3}\right)\right]^2 - \left[1+\ln(1-d)\right]^2 & \frac{2}{3} - e^{-3/2} \le d < 1 - e^{-1} \\ 1-\frac{2}{3}\left[\ln\left(d+\frac{1}{3}\right)\right]^2 & 1 - e^{-1} \le d < \frac{2}{3} \\ 1 & d \ge \frac{2}{3}, \end{cases}$$

which is plotted in Figure 13.12. Dots have been plotted at the breakpoints, with each of the lower four tightly-clustered breakpoints from Table 13.4 corresponding to a horizontal plane intersecting one of the four corners of region E in Figure 13.11. Percentiles of this distribution match the tabled values from Durbin [48]. We were not able to establish a pattern between the CDF of D_2 and the CDF of D_3 that might lead to a general expression for any n.

APPL was again used to calculate moments of D_3. The decimal approximations for the mean, variance, skewness, and kurtosis, are, respectively, $E(D_3) \cong 0.3727$, $V(D_3) \cong 0.008804$, $\gamma_3 \cong 0.4541$, and $\gamma_4 \cong 2.6538$. Although the functional form of the eight-segment PDF of D_3 is too lengthy to display here, it is plotted in Figure 13.13, with the only non-obvious breakpoint being on the initial nearly-vertical segment at $\left(d_{BCE}, f_{D_3}(d_{BCE})\right) \cong (0.2091, 1.5624)$.

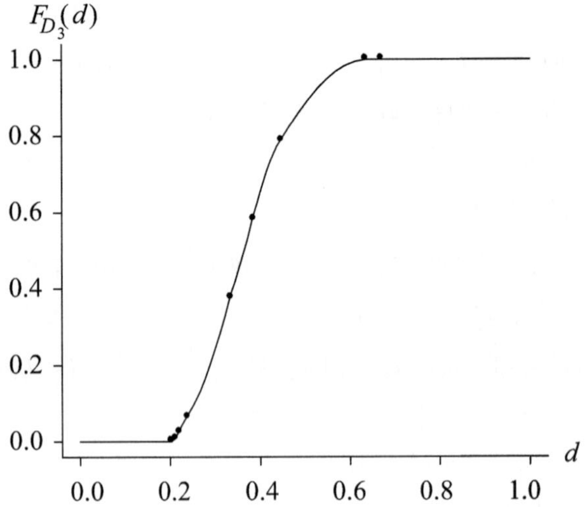

Fig. 13.12. The CDF of D_3

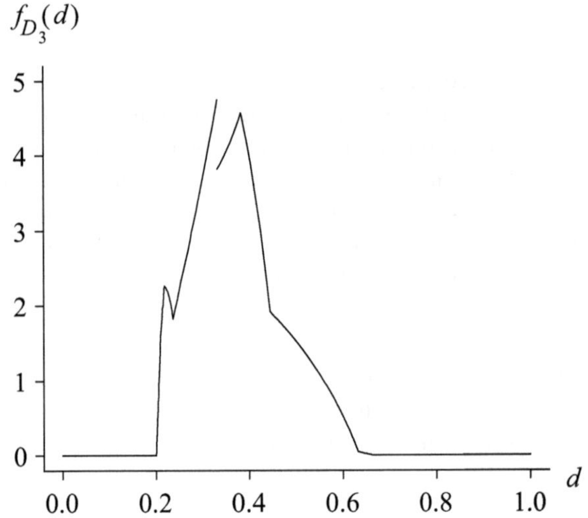

Fig. 13.13. The PDF of D_3

14

Parametric Model Discrimination for Heavily Censored Survival Data

A. Daniel Block and Lawrence M. Leemis

Abstract Simultaneous discrimination among various parametric lifetime models is an important step in the parametric analysis of survival data. We consider a plot of the skewness versus the coefficient of variation for the purpose of discriminating among parametric survival models. We extend the method of Cox and Oakes (1984, Analysis of Survival Data, Chapman & Hall/CRC) from complete to censored data by developing an algorithm based on a competing risks model and kernel function estimation. A by-product of this algorithm is a non-parametric survival function estimate.

Keywords Competing risks • Distribution selection • Kernel functions • Probability

---

Originally published in *IEEE Transaction on Reliability*, Volume 57, Number 2 in 2008, this paper makes extensive use of APPL from programming new ad hoc distributions to making new algorithms. Even the figures are produced by APPL. Of particular interest is the creation of the new procedure `CalcNetHaz` for calculating the net hazard function of the time to failure. Also APPL's `Mixture` command allows one to generate kernals from many distributions, so that non-parametric analyses are possible with the Cox and Oakes charts. Also in the appendix is an excellent example of a simulation using APPL code embedded in other Maple procedures. Also of interest is the instantiation of particular Weibull distributions and so forth in Figure 14.10, made possible with APPL coding.

A.D. Block
Block Consulting, Inc., Williamsburg, VA, USA

L.M. Leemis (✉)
The College of William and Mary, Williamsburg, VA, USA
e-mail: leemis@math.wm.edu

A.G. Glen, L.M. Leemis (eds.), *Computational Probability Applications*, International Series in Operations Research & Management Science 247,
DOI 10.1007/978-3-319-43317-2_14

14.1 Introduction

A well-known technique for choosing a probability model to approximate survival data is plot of the standardized third moment (skewness) versus the coefficient of variation as illustrated in Cox and Oakes [35]. Their plot gives the trajectory of the population skewness versus population coefficient of variation for several popular parametric lifetime models (e.g., Weibull and log-logistic). The sample skewness and the sample coefficient of variation can be plotted for a data set containing no censored data values. The proximity of the sample point to the population distribution values can be helpful in ruling in and ruling out particular parametric distributions as potential survival models.

This technique is superior to many of the standard exploratory data techniques for survival data (QQ plots, PP plots, probability plots, histograms, PDF estimation via kernel functions, empirical CDF estimation via the Kaplan–Meier product–limit estimate) in the sense that several competing models are easily viewed simultaneously. The weaknesses of this technique are (a) it considers only second and third moments which, for example, might do a poor job of detecting appropriate tail behavior, and (b) it is not easily adaptable to right-censored data sets. The focus of this paper is overcoming the second weakness. We limit our discussion to survival models.

Cox and Oakes' technique is extended here to allow the calculation of these statistics in the presence of heavy right censoring. Censored data is common in reliability and survival analysis. Recent work in the area includes works from Jiang and Jardine [69], Jiang et al. [70], Kundu and Sarhan [84], Li and Fard [100], Park [128], Sarhan [144], Soliman [149], and Zhang et al. [176, 177]. While data sets with light censoring (as is Cox and Oakes' example with the cancer data from Boag [18]) may be handled in a variety of ways (e.g., doubling censoring times or treating them as failures) without introducing significant error, heavily censored data leaves a data analyst with only heuristic methods that will significantly influence the position of the sample point. In this paper, we will present an analytical approach for plotting a point on the Cox and Oakes graph regardless of the fraction of censored observations in the data set.

Our approach assumes a random censoring scheme, in which the censoring times and the failure times are independent random variables. Treating censoring and failing as risks allows us to analyze the data set using a competing risks framework. We then treat the censored observations as coming from one distribution, and the failures as coming from a second distribution. We use kernel estimation to create empirical PDFs for these two data sets. Using a mathematically tractable distribution for the PDF, the competing risks model is used to estimate the failure time distribution as if censoring were not present. Because the failed and censored data are observed in the presence of another risk, these random variables represent crude lifetimes. A crude lifetime, as defined in David and Moeschberger [42], is the lifetime of an observation in the presence of other risks (also called a cause-specific distribution or cause-specific lifetime). Competing risks models are surveyed in more recently in Crowder [37] and Pintilie [132]. The goal of this paper is to find an estimate of the net lifetimes of the failure time distribution (i.e., when no censoring is present). The net lifetime is the lifetime of an observation only at risk of failing by one cause. We will use the crude lifetimes and statistical methods to "eliminate" the presence of the censoring risk. The result will be the distribution of the net lifetime of the time to failure which can then be treated as any other distribution of observed failures. The skewness and coefficient of variation associated with this distribution will then be plotted on the graph. The proximity of this point to curves associated with parametric survival models can be used to provide a list of appropriate models for data fitting purposes.

14.2 Literature Review

Since several methodologies are used in the algorithm that plots the sample skewness versus the sample coefficient of variation, we discuss their literature in the following subsections: (a) Cox and Oakes' methodology, (b) kernel functions, (c) competing risks.

Cox and Oakes' Methodology. Section 2.4 of Cox and Oakes [35] outlines four methods for plotting or tabulating data to select a parametric survival distribution. One of these methods (illustrated on page 27) is a plot of the standardized third moment (skewness) $\gamma_3 = E[(\frac{T-\mu}{\sigma})^3]$ versus the coefficient of variation $\gamma = \sigma/\mu$, where μ and σ are the mean and standard deviation of the lifetime T. Cox and Oakes plot γ_3 versus γ for several popular distributions. Their graph is replicated in Figure 14.1 using APPL code given in Appendix 1. The code has been modified and augmented from Evans and Leemis [51].

The exponential distribution occurs at the intersection of the Weibull and gamma curves at $\gamma = 1$ and $\gamma_3 = 2$. As stated earlier, Cox and Oakes "reasonably extrapolate" the light censoring in their example data set. Had the censoring been heavier, however, their analysis would have been unable to attain objective results. The square, cross, and diamond in Figure 14.1 are from the heavily-censored 6-MP treatment group from Gehan [56]:

$$6, 6, 6, 6^*, 7, 9^*, 10, 10^*, 11^*, 13, 16, 17^*, 19^*, 20^*, 22, 23, 25^*, 32^*, 32^*, 34^*, 35^*,$$

where an asterisk indicates a censored value. In this data set, there are $n = 21$ individuals at risk, 12 of which are right-censored. The points on the graph, starting from the left, treat the censored observations heuristically as follows:

1. Censored observations are treated as failures (square).
2. Censored observations are doubled, then treated as failures (diamond).
3. Censored observations are quadrupled, then treated as failures (cross).

The considerable scatter in these three points gives sufficient impetus to search for a parameter-free technique for plotting skewness versus coefficient of variation for a given data set. A goal of this paper is to devise an algorithm based on non parametric methods that will handle randomly right-censored data and not require parameters from an analyst.

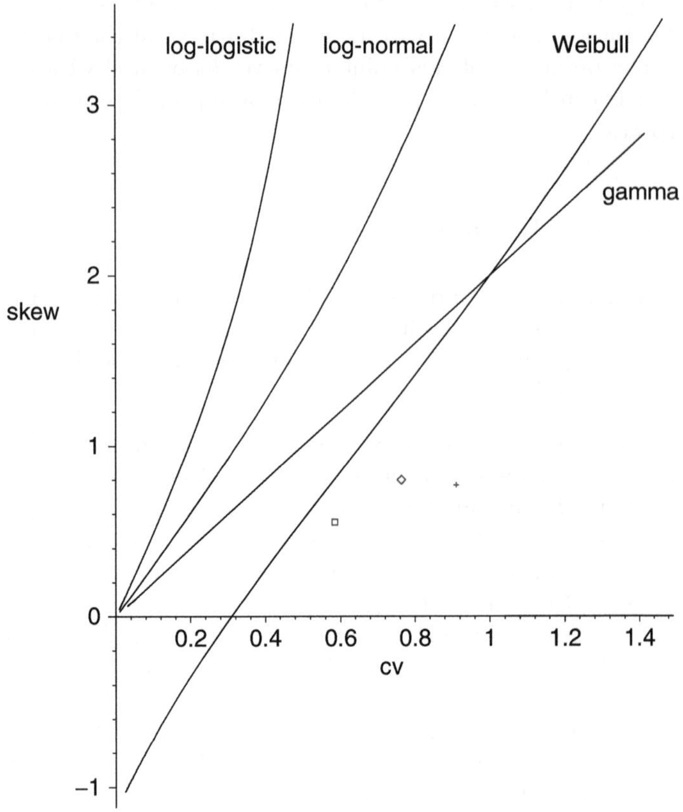

Fig. 14.1. The Cox and Oakes graph with three heuristic methods

Kernel Functions/Density Estimation. Bowman and Azzalini [20] give an excellent introductory framework for density estimation. They define a kernel density estimate as:

$$\hat{f}(y) = \frac{1}{n} \sum_{i=1}^{n} w(y - y_i; h),$$

where w, known as a kernel function, is a probability PDF whose variance is controlled by the smoothing parameter h and $y_1, y_2, \ldots, y_n$ are the data values.

The primary issue then, becomes the choice of the kernel function. For their example kernel function, Bowman and Azzalini use the normal distribution at $y - y_i$ with mean 0 and standard deviation h.

Section 3.4 of Silverman [147] gives comprehensive coverage on the topic of choosing the smoothing parameter h. Since density estimation is not a central topic of this paper, we will choose an h for use in the remainder of this work. Silverman gives two "quick ways of choosing" the smoothing parameter based on the properties of the sample data. The first is $h_1 = 0.79Rn^{-1/5}$, where R is the sample interquartile range, and the second is $h_2 = 0.9An^{-1/5}$, where $A = \min\{S, R/1.34\}$ and S is the sample standard deviation. Based on the discussion in Silverman, we will choose h_2, although this choice will have little influence on our results. Biswas and Sundaram [14] apply kernels functions to survival analysis.

Competing Risks. Treatment of censoring as a risk for use in a competing risks model is mentioned in Williams and Lagakos [168], Prentice and Kalbfleisch [134], David and Moeschberger [42], and Kalbfleisch and MacKay [74]. The articles of primary interest involve risk elimination and net/crude relationships.

Williams and Lagakos [168] address a common assumption in many censoring models that states that the survival mechanism and the censoring mechanism are independent. They give examples of cases in which this is an invalid assumption, and then examine the consequences of survival times and censoring times being dependent. They also discuss the testability of a data set for censoring influences on the survival mechanism. Where applicable, they give statistical tests that may be performed. Framed in the context of this paper, they discuss situations in which the crude lives of the observed data failing from different causes are dependent. Here, however, we will assume that the remaining longevity of an observation has no influence on its "surviving" censoring.

Prentice and Kalbfleisch [134] address three main issues: influence of regression coefficients on some, but not all failure causes, the interrelation between failure types, and the estimation of failure rates for one cause given the removal of some or all of the other causes. Starting from the assumption that the cause-specific hazard function is the only truly estimable portion of the competing risks model, they build models to address these three issues. While the first two issues are not relevant to this paper, the third is. They raise a list of concerns regarding cause-removal methods, and write about the most important concern: "In effect, the stochastic mechanism generating failures is assumed to continue beyond latent failure times for causes that have been removed until the smallest operative failure time is reached." The authors do not present a quantitative model. Their concerns were raised in response to Chiang [29] and do not effect our technique. Finally, the authors deal with censoring as a cause of failure. They abandon their previous concerns of cause removal stating, "the marginal distribution that arises from the elimination of censoring is clearly the relevant target of estimation." This is, of course, the very goal of this paper. They do, however, cite Williams and Lagakos [168] and raise the issue of the independence of survival and censoring—an issue we will avoid.

Kalbfleisch and MacKay [74] extend the work of Williams and Lagakos [168] and show that the constant-sum condition is equivalent to a simple relationship between hazard functions. David and Moeschberger [42] formulate a mathematical model for competing risks in their monograph. They mention the notion of treating censored data as a risk in their first chapter. Their main work, as it pertains to this paper, is in the area of net versus crude lifetimes. Leemis [93] also discusses the relationships between net and crude lifetimes. Section 5.1 contains an equation that will be central in making the analytical connection between net and crude lifetimes. This equation is proved in Leemis' Appendix 3 and states that, under the assumption of independent risks,

$$h_{Y_j}(t) = \frac{\pi_j f_{X_j}(t)}{\sum_{i=1}^{k} \pi_i S_{X_i}(t)}, \tag{14.1}$$

where Y_j denotes the net lifetime associated with risk j, X_j denotes the crude lifetime associated with risk j, π_j is the probability of failure from risk j, and f, S, and h denote the probability density, survivor, and hazard functions, respectively. The details on computing $h_{Y_j}(t)$ are given in Appendix 2. More recent references in the competing risks literature are given by Bousquet et al. [19].

14.3 A Parametric Example

Although the main emphasis of the work here is non parametric, the following parametric example demonstrates the calculation of the net lifetime of a failure distribution. For this example, we make the assumption that both the failure times and censoring times are exponentially distributed. That is, both failure times and censoring times are independent exponential random variables with different rates. For illustration, we will turn to the 6-MP treatment group data set listed earlier and found in Gehan [56]. Using competing risks terminology, risk 1 corresponds to censoring and risk 2 corresponds to failure. In the 6-MP data set, there are $n = 21$ patients on test and the number of observed failures (leukemia remissions) is $r = 9$. Using maximum likelihood, the failure rate associated with the first crude lifetime is

$$\hat{\lambda}_{X_1} = \frac{n - r}{\sum_{i|\delta_i=0} x_i} = \frac{12}{250},$$

where $\delta_i = 0$ denotes a censored observation, $x_i = \min\{c_i, t_i\}$, c_i is the censoring time for patient i and t_i is the failure time for patient i, for $i = 1, 2, \ldots, n$. The failure rate associated with the second crude lifetime is

$$\hat{\lambda}_{X_2} = \frac{r}{\sum_{i|\delta_i=1} x_i} = \frac{9}{109},$$

where $\delta_i = 1$ denotes a observed failure ($i = 1, 2, \ldots, n$). We estimate π_1 and π_2 as

$$\hat{\pi}_1 = \frac{n-r}{n} = \frac{12}{21} \qquad \text{and} \qquad \hat{\pi}_2 = \frac{r}{n} = \frac{9}{21}.$$

Because of our exponential assumption:

$$\hat{f}_{X_1}(t) = \hat{\lambda}_{X_1} e^{-\hat{\lambda}_{X_1} t}, \quad \hat{S}_{X_1}(t) = e^{-\hat{\lambda}_{X_1} t}, \quad \hat{f}_{X_2}(t) = \hat{\lambda}_{X_2} e^{-\hat{\lambda}_{X_2} t}, \quad \hat{S}_{X_2}(t) = e^{-\hat{\lambda}_{X_2} t},$$

for $t > 0$. Using Eq. (14.1) we compose the hazard function for the net lifetime of the failure data:

$$h_{Y_2}(t) = \frac{\pi_2 f_{X_2}(t)}{\pi_1 S_{X_1}(t) + \pi_2 S_{X_2}(t)},$$

for $t > 0$, which is estimated by

$$\hat{h}_{Y_2}(t) = \frac{\hat{\pi}_2 \hat{\lambda}_{X_2} e^{-\hat{\lambda}_{X_2} t}}{\hat{\pi}_1 e^{-\hat{\lambda}_{X_1} t} + \hat{\pi}_2 e^{-\hat{\lambda}_{X_2} t}} = \frac{\frac{27}{109} e^{-9t/109}}{4e^{-12t/250} + 3e^{-9t/109}},$$

for $t > 0$. The PDF of the net lifetime associated with failure is

$$f_{Y_2}(t) = h_{Y_2}(t) e^{-\int_0^t h_{Y_2}(\tau) d\tau},$$

and is estimated by

$$\hat{f}_{Y_2}(t) = \hat{h}_{Y_2}(t) e^{-\int_0^t \hat{h}_{Y_2}(\tau) d\tau}$$

for $t > 0$. This estimated PDF of the net lifetime of the failure data can be used to calculate the skewness and coefficient of variation of the failure data. These statistics are $(\hat{\gamma}, \hat{\gamma}_3) = (1.40, 2.55)$. In the remainder of the paper, we develop a non parametric method in order to avoid the assumption of exponentiality for the failure time and the censoring mechanism. Computation will be greatly simplified with the help of APPL.

14.4 Methodology

This section details the steps involved in extracting a survival distribution estimate from the observed failure and censoring times. We will use the following eight-point data set in our examples:

$$x = [0.25^*, 0.25, 0.35^*, 0.45^*, 1.00, 1.15^*, 1.25, 1.35],$$

where an asterisk indicates a right-censored observation. The data could also be described as:

$$x = [0.25, 0.25, 0.35, 0.45, 1.00, 1.15, 1.25, 1.35],$$

$$\delta = [0, 1, 0, 0, 1, 0, 1, 1],$$

where $x_i = \min\{t_i, c_i\}$, t_i is the failure time, c_i is the censoring time, δ_i is 1 when $x_i = t_i$ and 0 when $x_i = c_i$ for $i = 1, 2, \ldots, 8$. The random variables for failure times and censoring times are assumed to be independent. To verify the correctness of our method and the `CalcNetHaz` algorithm, we will work through two small examples.

The normal distribution was an obvious choice for a kernel function. There are, however, several issues that preclude its use. The first issue is the range of support. Establishing a normal kernel around any data value will include negative support values—an impossibility in lifetime data analysis because negative lifetimes never occur. The second issue is the intractability of the CDF. While APPL was able to return and plot a (rather complicated) CDF, it was unable to calculate the coefficient of variation or the skewness.

We will instead use the PDFs of the uniform and triangular distributions as kernel functions. These distributions allow for simple, tractable, CDFs and allow us to exploit APPL's piecewise function processing capability. We have avoided the problem of negative lifetimes with these distributions by carefully choosing our example data values and h so as to avoid negative support for the kernel function.

14.4.1 Uniform Kernel Function

Our method treats censored and observed data as coming from two different distributions. For simplicity in formulating, implementing and testing our algorithm we split the data into [0.25, 0.35, 0.45, 1.15] and [0.25, 1.0, 1.25, 1.35] where the first list contains the censoring times and the second list contains the failure times. These data values were chosen arbitrarily with preference given to ease of verification by hand. According to the chosen method for calculating bin width discussed in Silverman [147], we calculate h as $h = 0.9sn^{-1/5}$, where s is the sample standard deviation of the four values. For the first list, $h = 0.241$ and for the second, $h = 0.294$. We will, however, simplify by using $h = 0.25$ for our calculations in order to ease arithmetic and simplify verification. The first step is to create the two random variables with APPL:

```
> h := 0.25;
> C1 := UniformRV(0.25 - h, 0.25 + h);
> C2 := UniformRV(0.35 - h, 0.35 + h);
> C3 := UniformRV(0.45 - h, 0.45 + h);
> C4 := UniformRV(1.15 - h, 1.15 + h);
> X1 := Mixture([1 / 4, 1 / 4, 1 / 4, 1 / 4], [C1, C2, C3, C4]);
> F1 := UniformRV(0.25 - h, 0.25 + h);
> F2 := UniformRV(1.0 - h, 1.0 + h);
> F3 := UniformRV(1.25 - h, 1.25 + h);
> F4 := UniformRV(1.35 - h, 1.35 + h);
> X2 := Mixture([1 / 4, 1 / 4, 1 / 4, 1 / 4], [F1, F2, F3, F4]);
```

At this point, APPL has created two PDFs, $\hat{f}_{X_1}(t)$ and $\hat{f}_{X_2}(t)$. The kernel functions for these PDFs are uniformly distributed and centered at the observation with a width of $2h$. This APPL code returns the kernel function estimate for the crude censoring time PDF as

$$\hat{f}_{X_1}(t) = \begin{cases} 1/2 & 0 < t < 0.1 \\ 1 & 0.1 < t < 0.2 \\ 3/2 & 0.2 < t < 0.5 \\ 1 & 0.5 < t < 0.6 \\ 1/2 & 0.6 < t < 0.7 \\ 0 & 0.7 < t < 0.9 \\ 1/2 & 0.9 < t < 1.4. \end{cases}$$

Similarly, this APPL code returns the kernel function estimate for the crude failure time as

$$\hat{f}_{X_2}(t) = \begin{cases} 1/2 & 0 < t < 0.5 \\ 0 & 0.5 < t < 0.75 \\ 1/2 & 0.75 < t < 1.0 \\ 1 & 1.0 < t < 1.1 \\ 3/2 & 1.1 < t < 1.25 \\ 1 & 1.25 < t < 1.5 \\ 1/2 & 1.5 < t < 1.6. \end{cases}$$

Unless otherwise noted, all of the following plots were created using APPL's `PlotDist` command. Figure 14.2 is a plot of $\hat{f}_{X_1}(t)$ and Figure 14.3 is a plot of $\hat{f}_{X_2}(t)$.

In order to call the `CalcNetHaz` procedure, we must first compute a mixture of these PDFs. This mixture will serve as the denominator $\pi_1 S_{X_1}(t) + \pi_2 S_{X_2}(t)$ in the calculation of $h_{Y_2}(t)$. The π_i values come from the competing risks discussion and are the probability of failing due to the ith risk. In our example, four of the observations "fail" by censoring and the other four are observed failures. Therefore, $\hat{\pi}_1 = \hat{\pi}_2 = 0.5$. In general, for a data set with n items on test and r observed failures, $\hat{\pi}_1 = \frac{n-r}{n}$ and $\hat{\pi}_2 = \frac{r}{n}$. With APPL's `Mixture` procedure, we compute this denominator as a mixture of PDFs and convert it to a survivor function before passing it to `CalcNetHaz`.

```
> X12 := Mixture([1 / 2, 1 / 2], [X1, X2]);
```

This PDF $\hat{f}_{X_{12}}(t)$ has the following mathematical form and is plotted in Figure 14.4:

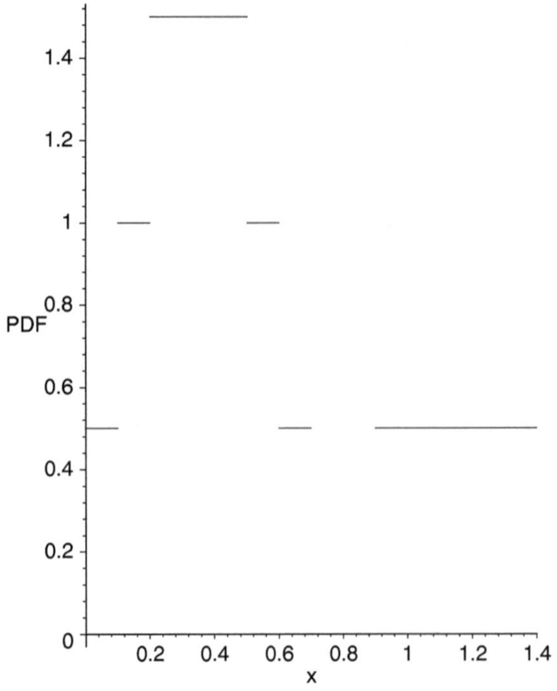

Fig. 14.2. Uniform kernel density estimate for the crude censoring time X_1

$$\hat{f}_{X_{12}}(t) = 0.5\hat{f}_{X_1}(t) + 0.5\hat{f}_{X_2}(t) = \begin{cases} 1/2 & 0 < t < 0.1 \\ 3/4 & 0.1 < t < 0.2 \\ 1 & 0.2 < t < 0.5 \\ 1/2 & 0.5 < t < 0.6 \\ 1/4 & 0.6 < t < 0.7 \\ 0 & 0.7 < t < 0.75 \\ 1/4 & 0.75 < t < 0.9 \\ 1/2 & 0.9 < t < 1.0 \\ 3/4 & 1.0 < t < 1.1 \\ 1 & 1.1 < t < 1.25 \\ 3/4 & 1.25 < t < 1.4 \\ 1/2 & 1.4 < t < 1.5 \\ 1/4 & 1.5 < t < 1.6. \end{cases}$$

Referring back to Eq. (14.1) to calculate a net lifetime from the crude lifetimes, the hazard function estimate for the net lifetime Y_2 associated with the crude lifetime X_2 is:

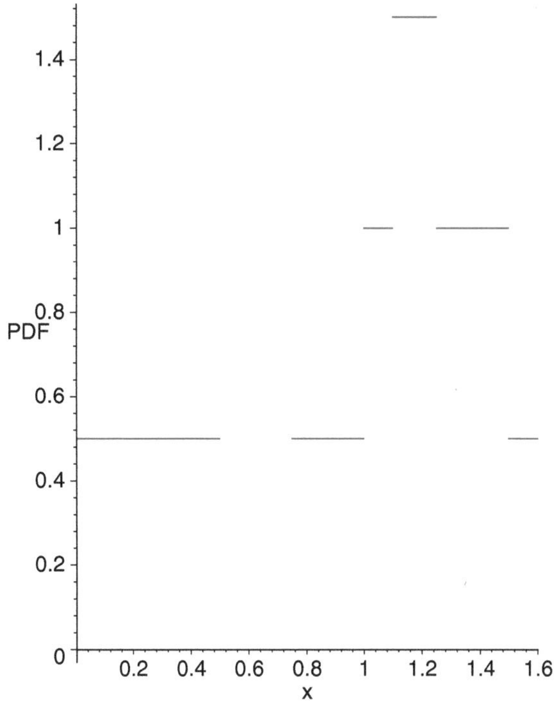

Fig. 14.3. Uniform kernel density estimate for the crude failure time X_2

$$h_{Y_2}(t) = \frac{\pi_2 f_{X_2}(t)}{\pi_1 S_{X_1}(t) + \pi_2 S_{X_2}(t)}.$$

The denominator has been calculated in Maple in its density form and can be converted to a survivor function with the APPL function `SF`. A call to `CalcNetHaz` with these three parameters gives the net lifetime hazard function for the observed failures.

```
> Y2 := CalcNetHaz(X2, SF(X12), 0.5);
```

Because X is the crude lifetime distribution and Y is the net lifetime distribution, we now have

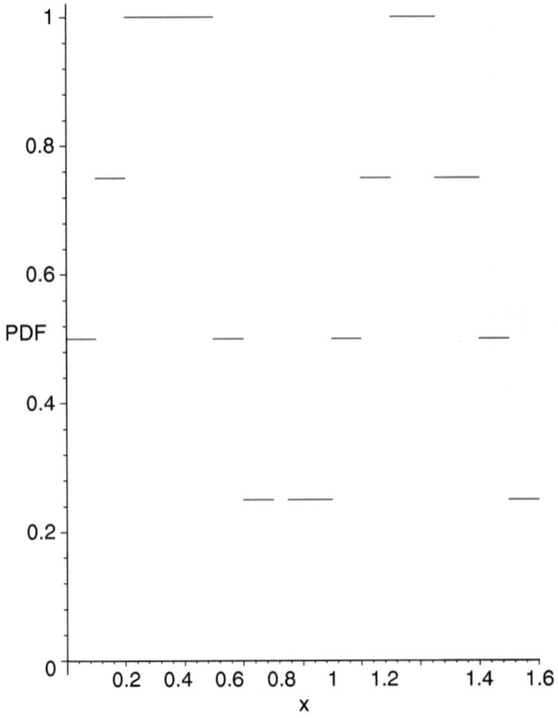

Fig. 14.4. Uniform kernel density estimate for X_{12}

$$\hat{h}_{Y_2}(t) = \frac{\pi_2 \hat{f}_{X_2}(t)}{\pi_1 \hat{S}_{X_1}(t) + \pi_2 \hat{S}_{X_2}(t)} = \begin{cases} \frac{1}{4-2t} & 0 < t < 0.1 \\ \frac{10}{41-30t} & 0.1 < t < 0.2 \\ \frac{10}{43-40t} & 0.2 < t < 0.5 \\ 0 & 0.5 < t < 0.75 \\ \frac{4}{11-4t} & 0.75 < t < 0.9 \\ \frac{20}{73-40t} & 0.9 < t < 1.0 \\ \frac{40}{3(31-20t)} & 1.0 < t < 1.1 \\ \frac{12}{23-16t} & 1.1 < t < 1.25 \\ \frac{4}{3(3-2t)} & 1.25 < t < 1.4 \\ \frac{20}{31-20t} & 1.4 < t < 1.5 \\ \frac{5}{8-5t} & 1.5 < t < 1.6. \end{cases}$$

The functional form of this hazard function, as computed by `CalcNetHaz`, has been verified by hand. Figure 14.5 contains a plot of $\hat{h}_{Y_2}(t)$, which has a vertical asymptote at $t = 1.6$. The following additional APPL statements give the coordinates of a point that can be plotted on Cox and Oakes' parametric model discrimination plot:

```
> cv := CoefOfVar(Y2);
> skew := Skewness(Y2);
```

These statements yield the point $(\hat{\gamma}, \hat{\gamma_3}) = (0.3608, -1.2320)$. When this same technique is applied to the 6-MP treatment group data with $h_2 = 5.798$ for the censored data and $h_2 = 3.970$ for the failure data, the point obtained is $(\hat{\gamma}, \hat{\gamma_3}) = (0.7338, 0.0425)$. Table 14.1 compares this point with the three points plotted in Figure 14.1. There is considerable difference between the heuristic approaches, the parametric analysis, and the competing risks approach.

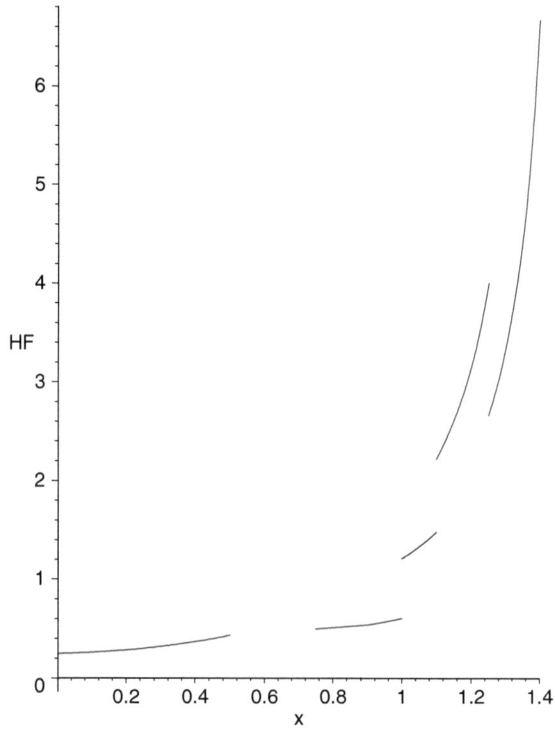

Fig. 14.5. Uniform kernel hazard function estimate for net lifetime of observed failures Y_2

14.4.2 Triangular Kernel Function

This section follows the technique and data set from Section 14.4.1 but uses the triangular distribution for the kernel function in the density estimation

Table 14.1. $\hat{\gamma}$ and $\hat{\gamma}_3$ for different methods for handling censoring

Method	$\hat{\gamma}$	$\hat{\gamma}_3$
Increase censoring times by 0 %	0.58	0.55
Increase censoring times by 50 %	0.69	0.76
Increase censoring times by 100 %	0.76	0.80
Parametric (exponential) analysis	1.40	2.55
Competing risks (uniform kernel)	0.73	0.04

method. The APPL code the calculating the kernel estimate for the PDFs of the crude lifetimes is:

```
> h := 0.25;
> C1 := TriangularRV(0.25 - h, 0.25, 0.25 + h);
> C2 := TriangularRV(0.35 - h, 0.35, 0.35 + h);
> C3 := TriangularRV(0.45 - h, 0.45, 0.45 + h);
> C4 := TriangularRV(1.15 - h, 1.15, 1.15 + h);
> X1 := Mixture([1 / 4, 1 / 4, 1 / 4, 1 / 4], [C1, C2, C3, C4]);
> F1 := TriangularRV(0.25 - h, 0.25, 0.25 + h);
> F2 := TriangularRV(1.0 - h, 1.0, 1.0 + h);
> F3 := TriangularRV(1.25 - h, 1.25, 1.25 + h);
> F4 := TriangularRV(1.35 - h, 1.35, 1.35 + h);
> X2 := Mixture([1 / 4, 1 / 4, 1 / 4, 1 / 4], [F1, F2, F3, F4]);
```

The three parameters in `TriangularRV` are the minimum, mode, and maximum. Again, $\hat{f}_{X_1}(t)$ and $\hat{f}_{X_2}(t)$ are the PDF on estimates. Their kernel functions are now the triangular distributions centered at the observation and having a width of $2h$.

$$\hat{f}_{X_1}(t) = \begin{cases} 4t & 0 < t < 0.1 \\ 8t - \frac{2}{5} & 0.1 < t < 0.2 \\ 12t - \frac{6}{5} & 0.2 < t < 0.25 \\ 4t + \frac{4}{5} & 0.25 < t < 0.35 \\ -4t + \frac{18}{5} & 0.35 < t < 0.45 \\ -12t + \frac{36}{5} & 0.45 < t < 0.5 \\ -8t + \frac{26}{5} & 0.5 < t < 0.6 \\ -4t + \frac{14}{5} & 0.6 < t < 0.7 \\ 0 & 0.7 < t < 0.9 \\ 4t - \frac{18}{5} & 0.9 < t < 1.15 \\ -4t + \frac{28}{5} & 1.15 < t < 1.4. \end{cases}$$

$$\hat{f}_{X_2}(t) = \begin{cases} 4t & 0 < t < 0.25 \\ 2 - 4t & 0.25 < t < 0.5 \\ 0 & 0.5 < t < 0.75 \\ 4t - 3 & 0.75 < t < 1.0 \\ 1 & 1.0 < t < 1.10 \\ 4t - \frac{17}{5} & 1.10 < t < 1.25 \\ \frac{8}{5} & 1.25 < t < 1.35 \\ \frac{62}{5} - 8t & 1.35 < t < 1.5 \\ \frac{32}{5} - 4t & 1.5 < t < 1.6. \end{cases}$$

These PDF estimates are plotted in Figures. 14.6 and 14.7.

As in the uniform case, we compute a mixture of these density functions to be the denominator of Eq. (14.1).

```
> X12 := Mixture([1 / 2, 1 / 2], [X1, X2]);
```

The PDF $\hat{f}_{X_{12}}(t)$ has the following form and is plotted in Figure 14.8:

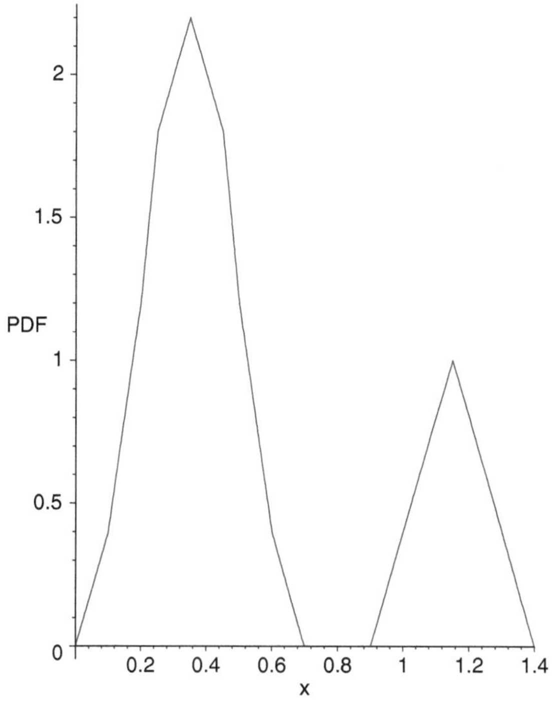

Fig. 14.6. Triangular kernel density estimate for crude censoring time X_1

$$\hat{f}_{X_{12}}(t) = 0.5\hat{f}_{X_1}(t) + 0.5\hat{f}_{X_2}(t) = \begin{cases} 4t & 0 < t < 0.1 \\ 6t - \frac{1}{5} & 0.1 < t < 0.2 \\ 8t - \frac{3}{5} & 0.2 < t < 0.25 \\ \frac{7}{5} & 0.25 < t < 0.35 \\ -4t + \frac{14}{5} & 0.35 < t < 0.45 \\ -8t + \frac{23}{5} & 0.45 < t < 0.5 \\ -4t + \frac{13}{5} & 0.5 < t < 0.6 \\ -2t + \frac{7}{5} & 0.6 < t < 0.7 \\ 0 & 0.7 < t < 0.75 \\ 2t - \frac{3}{2} & 0.75 < t < 0.9 \\ 4t - \frac{33}{10} & 0.9 < t < 1.0 \\ 2t - \frac{13}{10} & 1.0 < t < 1.1 \\ 4t - \frac{7}{2} & 1.1 < t < 1.15 \\ \frac{11}{10} & 1.15 < t < 1.25 \\ -2t + \frac{18}{5} & 1.25 < t < 1.35 \\ -6t + 9 & 1.35 < t < 1.4 \\ -4t + \frac{31}{5} & 1.4 < t < 1.5 \\ -2t + \frac{31}{5} & 1.5 < t < 1.6. \end{cases}$$

Using Eq. (14.1) as before, we want to calculate the hazard function of the net lifetime for the observed failures. A call to **CalcNetHaz** with these three parameters gives the hazard function for the net lifetimes for the observed failures.

```
> Y2 := CalcNetHaz(X2, SF(X12), 0.5);
```

If X is the crude lifetime distribution and Y is the net lifetime distribution for a random variable, the following hazard function results:

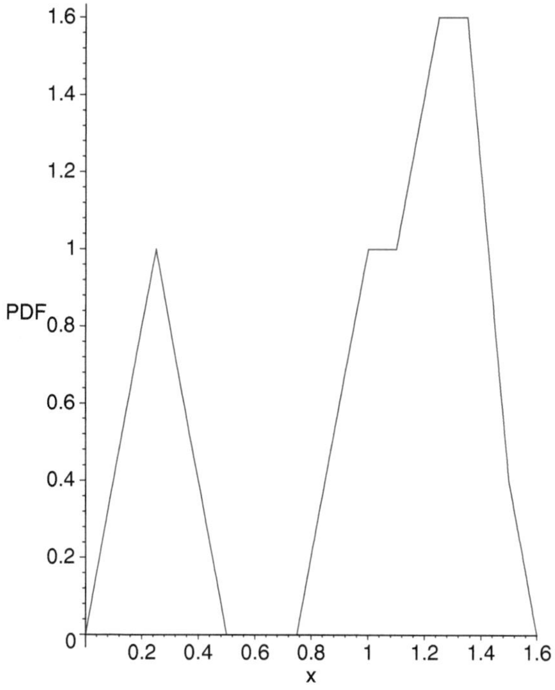

Fig. 14.7. Triangular kernel density estimate for crude failure time X_2

$$\hat{h}_{Y_2}(t) = \frac{\pi_2 \hat{f}_{X_2}(t)}{\pi_1 \hat{S}_{X_1}(t) + \pi_2 \hat{S}_{X_2}(t)} = \begin{cases} \frac{2t}{-2t^2+1} & 0 < t < 0.1 \\ \frac{200t}{99+20t-300t^2} & 0.1 < t < 0.2 \\ \frac{40t}{19+12t-80t^2} & 0.2 < t < 0.25 \\ \frac{5(2t-1)}{-6+7t} & 0.25 < t < 0.35 \\ \frac{-200(2t-1)}{289-560t+400t^2} & 0.35 < t < 0.45 \\ \frac{-20(2t-1)}{37-92t+80t^2} & 0.45 < t < 0.5 \\ 0 & 0.5 < t < 0.75 \\ \frac{-8(4t-3)}{1-24t+16t^2} & 0.75 < t < 0.9 \\ \frac{-200(4t-3)}{349-1320t+800t^2} & 0.9 < t < 1.0 \\ \frac{200}{51+520t-400t^2} & 1.0 < t < 1.1 \\ \frac{-40(-17+20t)}{433-1400t+800t^2} & 1.1 < t < 1.15 \\ \frac{8(-17+20t)}{125-88t} & 1.15 < t < 1.25 \\ \frac{32}{125-144t+40t^2} & 1.25 < t < 1.35 \\ \frac{-20(-31+20t)}{677-900t+300t^2} & 1.35 < t < 1.4 \\ \frac{-20(-31+20t)}{481-620t+200t^2} & 1.4 < t < 1.5 \\ \frac{10}{8-5t} & 1.5 < t < 1.6. \end{cases}$$

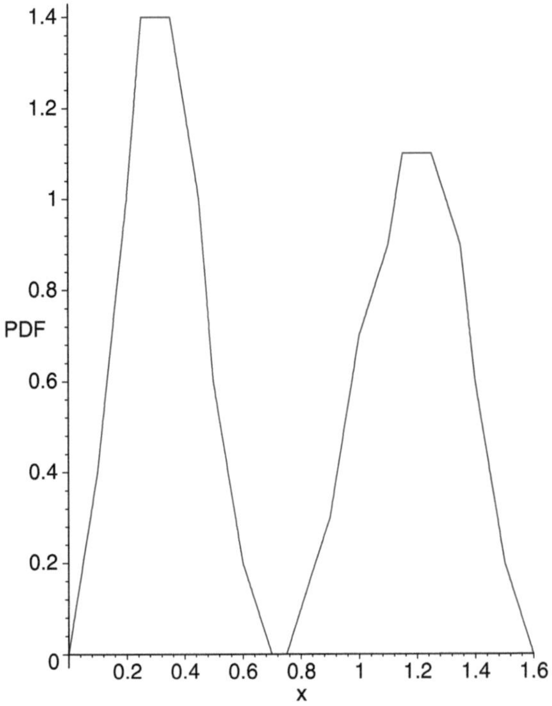

Fig. 14.8. Triangular kernel density estimate for X_{12}

The functional form of this hazard function has also been verified by hand. It is plotted in Figure 14.9. As in the uniform case, $\hat{h}_{Y_2}(t)$ has a vertical asymptote at 1.6. Using APPL, the coordinates of this distribution on Cox and Oakes' parametric model discrimination plot are calculated as follows:

```
> cv := CoefOfVar(Y2);
> skew := Skewness(Y2);
```

which yields the point $(\hat{\gamma}, \hat{\gamma_3}) = (0.3452, -1.3390)$. Not surprisingly, this point is in reasonable proximity to the point $(0.3608, -1.2320)$ obtained with the uniform kernel.

14.5 Monte Carlo Simulation Analysis

The Monte Carlo analysis of our algorithm begins with a plot of $\hat{\gamma_3}$ vs. $\hat{\gamma}$ for several parametric distributions which will serve as a baseline to assess how well our algorithm adapts to censored data values. The plot for various Weibull and log-logistic distributions appears in Figure 14.10. The points plotted are associated with a sample size of $n = 1000$ for each of four parametric distributions (one log-logistic and three Weibull with shape parameters 4, 2, and

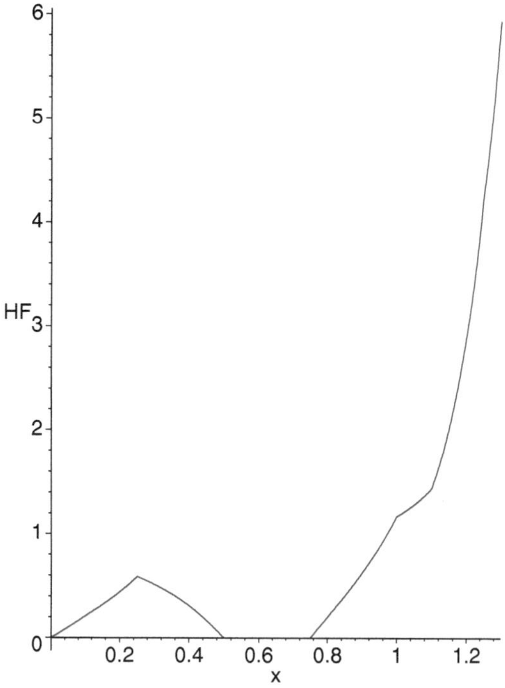

Fig. 14.9. Triangular kernel hazard function estimate

0.8 from left to right). The plotted values cluster around the correct point on the chart, but have considerable dispersion for such a large sample size. Not surprisingly, the dispersion associated with the skewness consistently exceeds the dispersion associated with the coefficient of variation. Also, as the shape parameter in the Weibull distribution decreases, the points spread. Some of the parameter choices exhibit a positive correlation between $\hat{\gamma}$ and $\hat{\gamma}_3$.

Appendix 3 contains a Maple implementation of a Monte Carlo simulation of the method described here. Figure 14.11 displays points on the Cox and Oakes' graph where the lifetimes are drawn from a Weibull population with a shape parameter of 5 and were calculated from data sets of $n = 200$ values. Computational time requirements (since Maple-based APPL is interpreted) prevented the plotting of as many values as in Figure 14.10. The points cluster around the appropriate point on the Weibull curve, although there are several points that fall significantly to the southeast of the target. The smaller number of items on test relative to the simulation illustrated in Figure 14.10 results in a wider dispersion of the plotted values.

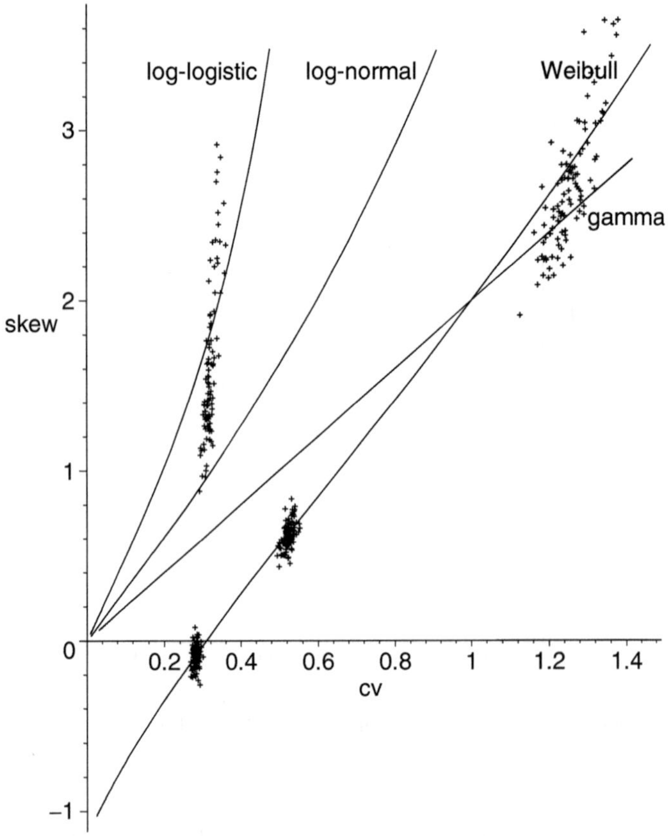

Fig. 14.10. Complete data sets for Weibull with $\kappa = 4, 2, 0.8$ and log-logistic with $\kappa = 6$

14.6 Conclusions and Further Work

Cox and Oakes' parametric survival model discrimination plot has been extended to the case of a right-censored survival data set using kernel functions to estimate PDFs of the crude lifetime PDFs and a competing risks framework to adjust for the effect of right-censoring. APPL and Maple were used for numerical and analytical computations as well as for bookkeeping piecewise functions.

Future work on this topic could proceed in several directions. Other kernel functions could be used to obtain a scatter-plot of skewness versus coefficient of variation. Using the Weibull distribution with $\kappa = \frac{1}{1-\ln(2)} \cong 3.25889$ (where the mode equals the median to give a bell-shaped kernel density function estimate) would be a reasonable approximation to the normal distribution and would prevent negative lifetime support. In addition, since the Weibull

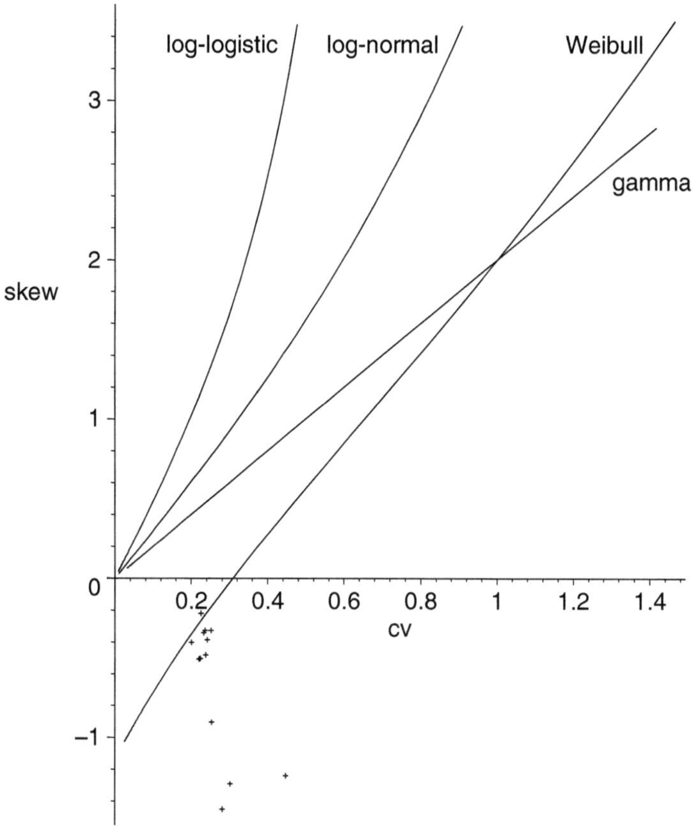

Fig. 14.11. Censored data sets for Weibull with shape parameter 5 with sample size $n = 200$

distribution has positive support, there will be fewer piecewise segments to account for. The influence that h has on a plotted point could also be investigated.

Appendix 1

This appendix contains the APPL code to create the Cox and Oakes' graph shown in Figure 14.1.

```
> unassign('kappa'):
> lambda := 1:
> X := GammaRV(lambda, kappa):
> c := CoefOfVar(X):
> s := Skewness(X):
> GammaPlot := plot([c, s, kappa = 0.5 .. 999], labels = ["cv",
```

```
>               " skew"]):
> unassign('kappa'):
> lambda := 1:
> X := WeibullRV(lambda, kappa):
> c := CoefOfVar(X):
> s := Skewness(X):
> WeibullPlot := plot([c, s, kappa = 0.7 .. 50.7]):
> unassign('kappa'):
> lambda := 1:
> Y := LogNormalRV(lambda, kappa):
> c := CoefOfVar(Y):
> s := Skewness(Y):
> LogNormalPlot := plot([c, s, kappa = 0.01 .. 0.775]):
> unassign('kappa'):
> lambda := 1:
> Y := LogLogisticRV(lambda, kappa):
> c := CoefOfVar(Y):
> s := Skewness(Y):
> LogLogisticPlot := plot([c, s, kappa = 4.3 .. 200.5]):
> cnsrgrp := plot([[0.5849304, 0.5531863]], style = point,
>                 symbol = box):
> cnsrgrp15 := plot([[0.6883908, 0.760566]], style = point,
>                 symbol = cross):
> cnsrgrp20 := plot([[0.7633863, 0.8009897]], style = point,
>                 symbol = diamond):
> with(plots):
> lll := textplot([0.17, 3.3, "log-logistic"], 'align =
>       {ABOVE, RIGHT}'):
> lnl := textplot([0.59, 3.3, "log-normal"], 'align =
>       {ABOVE, RIGHT}'):
> wbl := textplot([1.2, 3.3, "Weibull"], 'align =
>       {ABOVE, RIGHT}'):
> gml := textplot([1.3, 2.44, "gamma"], 'align =
>       {ABOVE, RIGHT}'):
> plots[display]({lll, lnl, wbl, gml, GammaPlot, WeibullPlot,
>       LogNormalPlot, LogLogisticPlot, cnsrgrp, cnsrgrp15,
>       cnsrgrp20}, scaling = unconstrained);
```

Appendix 2

Computational Issues. With our kernel density estimates being mixtures of large numbers of random variables, it became clear that even small data sets could result in piecewise functions with an unmanageable number of segments. To assist in the computation of these functions, we turned to APPL. In addition, APPL allows for the creation and combination of all types of standard random variables (uniform, normal, triangular, Weibull, etc.)—the very random variables we use in our kernel functions. The flexibility of APPL will allow for the efficient manipulation of many random variables.

Despite APPL's comprehensive random variable handling ability, the equation at the core of our analysis, Eq. (14.1) has not been implemented. This necessitated our devising an algorithm (using the APPL language as a platform) that could perform the implementation of Eq. (14.1) for random variables defined in a piecewise manner. The Maple function `CalcNetHaz` calculates $h_{Y_j}(t)$ for crude lifetimes defined in a piecewise manner. It must be passed the APPL PDF for the numerator $f_{X_j}(t)$, a mixture of APPL survival functions for the denominator, and the numerator's π_j values. The procedure `CalcNetHaz` returns the hazard function of the time to failure using Eq. (14.1). The code used to check for

- the correct number of arguments,
- the correct format for the PDF of the numerator and mixture of survivor functions in the denominator,
- the correct type (continuous) of random variables $X_1, X_2, \ldots, X_k$,
- the numerator given as a PDF and the denominator as a SF,
- $0 < \pi_j < 1$,

is suppressed for brevity. Since the kernel estimate for the failure and censoring distributions may be defined in a piecewise fashion (e.g., for a uniform or triangular kernel), the procedure accommodates piecewise distributions.

```
> CalcNetHaz := proc(num :: list(list), denom :: list(list),
>       NumPI :: float)
>    local retval, nsegn, i, j:
>    retval := []:
>    nsegn := nops(num[2]):
>    i := 1:
>    for j from 2 by 1 to nsegn do
>      while denom[2][i] < num[2][j] do
>        retval := [op(retval), unapply(simplify(
>                    (NumPI * num[1][j - 1])
>                    / denom[1][i])(x), x)]:
>        i := i + 1:
>      end do:
>    end do:
>    return([retval, denom[2], ["Continuous", "HF"]]):
> end:
```

The first two arguments to `CalcNetHaz` are lists of three lists, the last argument, π_j, is a scalar. The first list in the first parameter's three lists is of the numerator's $n-1$ PDFs. These correspond to the n breakpoints in the numerator. The first list in the second parameter's three lists is of the denominator's $m-1$ PDFs. These correspond to the m breakpoints in the denominator. These breakpoints are found in the second of the three lists. The third list contains the strings "Continuous" and either "PDF" or "SF" to denote the type of distribution representation. For each of the segments, the algorithm calculates a hazard function for the current segment based on Eq. (14.1). The algorithm assumes that the denominator is a mixture distribution involving the term in the numerator. This assumption can be made because in Eq. (14.1), since S_{X_j} in the denominator is derived from f_{X_j} in the numerator (or vice-versa) and results in denominator segment breaks that are a superset of those in the numerator. After looping through each of the segments, the algorithm returns the list of hazard functions along with the segment breaks of the denominator.

Appendix 3

This appendix contains the Monte Carlo simulation code in APPL necessary to conduct the experiments described in Section 14.5.

```
> n := 1000:
> kappa := 5:
> for i from 1 to 80 do
>   r := 0:
>   X1 := [ ]:
>   X2 := [ ]:
>   for k from 1 to n do
>     f := -log(UniformVariate()) ^ (1 / kappa):
>     c := -log(UniformVariate()) ^ (1 / kappa):
>     if f < c then
>       r := r + 1:
>       X2 := [op(X2), f]:
>     else
>       X1 := [op(X1), c]:
>     end if:
>   end do:
>   if r < n and r > 0 then
>     dists := [ ]:
>     weights := [ ]:
>     R := describe[quartile[3]](X1) - describe[quartile[1]](X1):
>     h := 0.79 * R * (n - r) ^ (-0.2):
>     for j from 1 to n - r do
>       weights := [op(weights), 1 / (n - r)]:
>       if h > sort(X1)[1] then
>         fd := fopen("mapsim2", APPEND):
>         fprintf(fd, "The following line was padded:
>                 h=%g min=%g\n", h, sort(X1)[1]):
>         fclose(fd):
>         h := sort(X1)[1]:
>       end if:
```

```
>       dists := [op(dists), UniformRV(X1[j] - h, X1[j] + h)]:
>     od:
>     f_X1 := Mixture(weights, dists):
>     dists := [ ]:
>     weights := [ ]:
>     R := describe[quartile[3]](X2) - describe[quartile[1]](X2):
>     h := 0.79 * R * (r ^ (-0.2)):
>     for j from 1 to r do
>       weights := [op(weights), 1 / r]:
>       if h > sort(X2)[1] then
>         fd := fopen("mapsim2", APPEND):
>         fprintf(fd, "The following line was padded: h=%g
>                 min=%g\n", h, sort(X2)[1]):
>         fclose(fd):
>         h := sort(X2)[1]:
>       end if:
>       dists := [op(dists), UniformRV(X2[j] - h, X2[j] + h)]:
>     od:
>     f_X2 := Mixture(weights, dists):
>     f_X12 := Mixture([(n - r) / n, r / n], [f_X1, f_X2]):
>     h_Y2 := CalcNetHaz(f_X2, SF(f_X12), evalf(r / n)):
>     f_Y2 := PDF(h_Y2):
>     mu := Mean(f_Y2):
>     ExpValueXSqrd := ExpectedValue(f_Y2, x -> x ^ 2):
>     sigma := sqrt(ExpValueXSqrd - mu ^ 2):
>     Term1 := ExpectedValue(f_Y2, x -> x ^ 3):
>     Term2 := 3 * mu * ExpValueXSqrd:
>     Term3 := 2 * mu ^ 3:
>     skew := (Term1 - Term2 + Term3) / sigma ^ 3:
>     cov := sigma / mu:
>     fd := fopen("mapsim2", APPEND):
>     fprintf(fd, "[[%g, %g]], \n", Re(cov), Re(skew)):
>     fclose(fd):
>   elif r = n then
>     skew := describe[skewness](X2):
>     cov := describe[coefficientofvariation](X2):
>     fd := fopen("mapsim4", APPEND):
>     fprintf(fd, "[[%g, %g]], \n", Re(cov), Re(skew)):
>     fclose(fd):
>   end if:
> end do:
```

15

Lower Confidence Bounds for System Reliability from Binary Failure Data Using Bootstrapping

Lawrence M. Leemis

Abstract Binary failure data are collected for each of the independent components in a coherent system. Bootstrapping is used to determine a $(1 - \alpha)100\,\%$ lower confidence bound on the system reliability. When a component with perfect test results is encountered, a beta prior distribution is used to avoid an overly optimistic lower bound.

Keywords Beta distribution • Binomial confidence interval • Coherent system • Computer algebra system

15.1 Introduction

We consider the problem of determining a $(1-\alpha)100\,\%$ lower confidence bound on the system reliability for a coherent system of k components using the failure data (y_i, n_i), where y_i is the number of components of type i that pass the test and n_i is the number of components of type i on test, $i = 1, 2, \ldots, k$. We assume throughout that the components fail independently, e.g., no common-cause failures. The outline of the article is as follows. We begin with the case of a single ($k = 1$) component system where n components are placed on a test

---

Originally published in the *Journal of Quality Technology*, Volume 38 in 2006, this work uses APPL as one of its technologies that enabled the research. The initial problem arose in a reliability problem posed by engineers. This is the first publication that mentions using APPL to eliminate resampling error of bootstrap studies. The code relies on the procedures `Transform` and `Product`, to in effect establish a bootstrap with $B = \infty$.

L.M. Leemis
The College of William and Mary, Williamsburg, VA, USA
e-mail: leemis@math.wm.edu

A.G. Glen, L.M. Leemis (eds.), *Computational Probability Applications*, International Series in Operations Research & Management Science 247,
DOI 10.1007/978-3-319-43317-2_15

and y components pass the test. The Clopper–Pearson lower bound is used to provide a lower bound on the reliability. This model is then generalized to the case of multiple ($k > 1$) components. Bootstrapping is used to estimate the lower confidence bound on system reliability. We then address a weakness in the bootstrapping approach—the fact that the sample size is moot in the case of perfect test results, e.g., when $y_i = n_i$ for some i. This weakness is overcome by using a beta prior distribution to model the component reliability before performing the bootstrapping. Two subsections consider methods for estimating the parameters in the beta prior distribution for components with perfect test results. The first subsection considers the case when previous test results are available, and the second subsection considers the case when no previous test results are available. A simulation study compares various algorithms for calculating a lower confidence bound on the system reliability. The last section contains conclusions.

15.2 Single-Component Systems

Single-component systems are considered first because (1) there are known approximate confidence intervals for the lower reliability confidence bound and (2) these intervals will be used later in the paper to help determine the appropriate parameters for the beta distribution in the case where no prior test results exist on the component of interest.

Let n components be placed on test and let y of these components pass the test. Under the assumption that the test values (1 for pass, 0 for failure) $X_1, X_2, \ldots, X_n$ are independent and identically distributed Bernoulli random variables with unknown parameter p, $Y = \sum_{i=1}^{n} X_i$ is a binomial random variable with parameters n and p. The maximum likelihood estimator for p is $\hat{p} = Y/n$, which is unbiased and consistent. The interest here is in a lower confidence bound for the reliability p.

There is a wide literature on confidence intervals of this type since a confidence interval on a proportion is of interest on anything from a political poll to consumer preference. Vollset [164] compares 13 confidence intervals and Newcombe [121] compares 7 confidence intervals. Rather than fine-tuning these intervals as has been suggested by many authors, we have settled on using the Clopper–Pearson (CP) "exact" interval even though Newcombe [121, page 201] points out that its status as a gold standard has been disputed recently because the method is conservative, i.e., the actual coverage is greater than or equal to the stated coverage (see Agresti and Coull [2] for details).

Let $p_L < p < p_U$ be an "exact" (see Blyth [17]) CP two-sided confidence interval for p, where p_L and p_U are functions of the sample size n, the number of successes y, and the stated coverage of the interval, $1 - \alpha$. This is an approximate confidence interval due to the discrete nature of the binomial distribution. For $y = 1, 2, \ldots, n - 1$ the lower limit p_L satisfies (see, for example, Agresti and Coull [2])

$$\sum_{k=y}^{n} \binom{n}{k} p_L^k (1 - p_L)^{n-k} = \alpha/2.$$

For $y = 1, 2, \ldots, n-1$, the upper limit p_U satisfies

$$\sum_{k=0}^{y} \binom{n}{k} p_U^k (1 - p_U)^{n-k} = \alpha/2.$$

As shown in Leemis and Trivedi [98], these confidence interval limits can be expressed in terms of quantiles of the F distribution:

$$\left(1 + \frac{n-y+1}{yF_{2y,2(n-y+1),1-\alpha/2}}\right)^{-1} < p < \left(1 + \frac{n-y}{(y+1)F_{2(y+1),2(n-y),\alpha/2}}\right)^{-1},$$

where the third subscript on F refers to the right-hand tail probability.

Simply reallocating the probability α to the lower limit gives the following lower confidence bound for the reliability:

$$p_L = \left(1 + \frac{n-y+1}{yF_{2y,2(n-y+1),1-\alpha}}\right)^{-1}$$

for $y = 1, 2, \ldots, n-1$. For the case of all failures ($y = 0$), the lower bound is, of course, $p_L = 0$. For the case of all passes ($y = n$), the lower bound is $p_L = \alpha^{1/n}$.

Example 15.1. For CP lower confidence interval bounds, we use the following four sets of values for n and y, which give point estimates and 95 % CP lower confidence interval bounds for the reliability:

$$\begin{aligned}
n &= 10, \quad y = 7 \quad \Rightarrow \hat{p} = 0.7, \quad p_L = 0.393.\\
n &= 100, y = 97 \quad \Rightarrow \hat{p} = 0.97, p_L = 0.924.\\
n &= 10, \quad y = 10 \quad \Rightarrow \hat{p} = 1.0, \quad p_L = 0.741.\\
n &= 100, y = 100 \Rightarrow \hat{p} = 1.04, p_L = 0.970.
\end{aligned}$$

An S-Plus function named `confintlower` is given in Appendix 1 which can be used to calculate these lower confidence interval bounds. Figure 15.1 is a plot of y vs. p_L when $n = 10$ for $\alpha = 0.10, 0.05, 0.01$, with the points connected with line segments. Figure 15.2 contains a similar plot for $n = 100$. The lower bounds are monotonic in y, n, and α.

15.3 Multiple-Component Systems

A three-component ($k = 3$) series system is used as an example throughout this section, although the techniques described here apply to any coherent

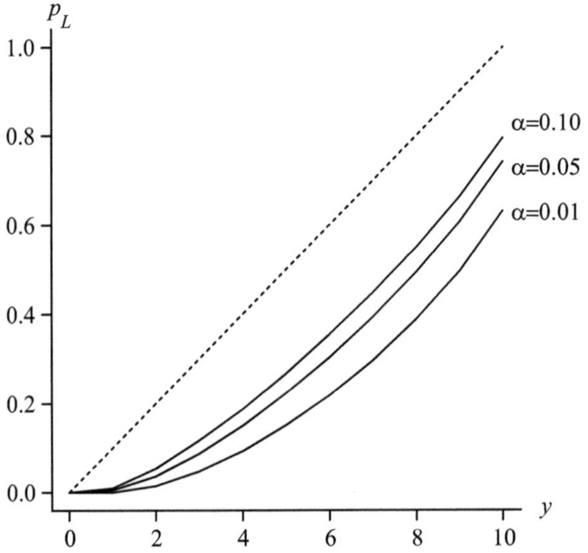

Fig. 15.1. Point estimate (*dashed*) and CP lower confidence bounds (*solid*) when $n = 10$

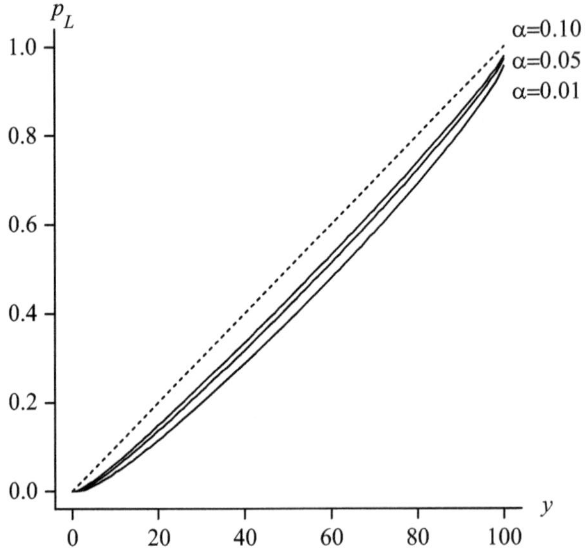

Fig. 15.2. Point estimate (*dashed*) and CP lower confidence bounds (*solid*) when $n = 100$

system of k independent components. The number of components tested and the number of passes for each type of component for the example are given in Table 15.1. The point estimate for the system reliability is

$$\frac{21}{23} \cdot \frac{27}{28} \cdot \frac{82}{84} = \frac{1107}{1288} \cong 0.8595.$$

The remainder of this section involves the use of bootstrapping (Efron and Tibshirani [49]) to calculate a lower 95 % confidence interval bound. Other authors (e.g., Martin [108] and Padgett and Tomlinson [126]) have used bootstrapping for determining confidence limits. The approach used here differs conceptually from the standard bootstrap problem, where the standard error of a *single* unknown distribution is estimated by resampling iid data. In our setting, there are k *different* distributions (one for each component) and we resample component reliabilities and combine using the reliability function to yield the system reliability.

Table 15.1. Failure data for a three-component series system example

Component number	$i = 1$	$i = 2$	$i = 3$
Number passing (y_i)	21	27	82
Number on test (n_i)	23	28	84

Bootstrapping resamples B of the systems, calculates the system reliability, then outputs the αB^{th} ordered system reliability. More specifically, for the three-component system of interest, the bootstrapping algorithm follows the following steps.

- For the first component, the data set (21 ones and 2 zeros) is sampled with replacement 23 times.
- These values are summed and divided by 23 yielding a reliability estimate for the first component.
- The previous two steps are repeated for components 2 and 3.
- The product of the reliability estimates for the three components are multiplied (because the components are arranged in series and their failures are independent) to give a system reliability estimate.

The above procedure is repeated B times. The B system reliability estimates are then sorted. Finally, the αB^{th} ordered system reliability is output, which is used as a lower bound on the system reliability.

The algorithm for estimating the $(1 - \alpha)100\,\%$ lower confidence interval bound is given in Table 15.2, where $\tilde{p}_i$ is a bootstrap estimate for the reliability of component i and z_j is a bootstrap estimate of the system reliability. The binomial distribution is appropriate because the resampling from the data set is performed with replacement. In the pseudocode in Table 15.2, indentation is used to indicate begin-end blocks. The returned value $z_{\alpha B}$ is the order statistic associated with the z_j's generated in the outside loop. See Law and Kelton [89] for handling the case when αB is not an integer.

This algorithm has been implemented in S-Plus as a function named `seriessystemboot` which is given in Appendix 2. The first two arguments,

Table 15.2. Bootstrap algorithm for calculating a $(1-\alpha)100\,\%$ lower confidence bound for the reliability of a k-component series system

for j from 1 to B	[resampling loop]
for i from 1 to k	[loop through components]
$\tilde{p}_i \leftarrow \text{Binomial}(n_i, y_i/n_i)/n_i$	[component i reliability]
$z_j \leftarrow \prod_{i=1}^{k} \tilde{p}_i$	[calculate system reliability]
sort $\mathbf{z}$	[sort the system reliability values]
return $z_{\alpha B}$	[return the estimate for the lower bound]

`n` and `y`, are vectors of length k, and the third argument, `alpha`, is a real number between 0 and 1, e.g.,

```
> seriessystemboot(c(23, 28, 84), c(21, 27, 82), 0.05)
```

prints a point estimate and a 95 % lower confidence interval bound on the system reliability for the three-component series system considered in this section. After a call to `set.seed(3)` to set the random number seed, five calls to `seriessystemboot` yield the following estimates for p_L:

$$0.7426057 \qquad 0.7482993 \qquad 0.7456744 \qquad 0.7486690 \qquad 0.7453416.$$

The dispersion associated with these five estimates is due to the finite choice of B, i.e., $B = 10{,}000$.

Resampling error can be eliminated using APPL. The APPL statements given in Appendix 3 utilize the `Product` and `Transform` procedures. This alternative approach to determining a lower 95 % bootstrap confidence interval bound for the system reliability is equivalent to using an infinite value for B. Since $\tilde{p}_i$ can assume any one of $n_i + 1$ values, there are a possible $24 \cdot 29 \cdot 85 = 59{,}160$ potential mass values for the random variable `T` determined by the `Product` procedure. Of these, only 6,633 are distinct since the `Product` procedure combines repeated values. Since the random variable `T` from the APPL code plays an analogous role to the vector $\mathbf{z}$ from the bootstrap algorithm in Table 15.2, the lower 95 % bootstrap confidence interval bound is the 0.05 fractile of the distribution of `T`, which is $p_L = 6723/9016 \cong 0.746$. This result using APPL is consistent with the standard resampling approach for finding the lower confidence interval limit based on the five results presented earlier (one equals 6723/9016 exactly, two fall above 6723/9016, and two fall below 6723/9016).

How Well Does the Bootstrap Procedure Perform? This question is difficult to address because there is no exact interval to compare with, even in the case of a single component. It is instructive, however, to isolate one component and compare the CP approach described in the previous section with bootstrapping. Arbitrarily choosing the second component with $n_2 = 28$ items on test, Figure 15.3 shows the CP lower confidence bound and the bootstrap lower confidence bound for $\alpha = 0.05$ and for $y_2 = 10, 11, \ldots, 28$. The bootstrap lower confidence interval limit does not require any iteration, since the value plotted is $q/28$, where q is the smallest integer that satisfies

$$\sum_{k=0}^{q} \binom{n_2}{k} \left(\frac{y_2}{28}\right)^k \left(1 - \left(\frac{y_2}{28}\right)\right)^{n_2-k} \geq \alpha,$$

i.e., q is the α-quantile of a binomial distribution with $n_2 = 28$ trials and probability of success $y_2/28$. Figure 15.3 shows that

- the bootstrap interval is more susceptible to the discrete nature of the binomial sampling scheme
- the CP interval is wider than the bootstrap interval.

Figure 15.3 also points out a glaring deficiency in the bootstrapping approach that was not revealed in the example in this section because all three of

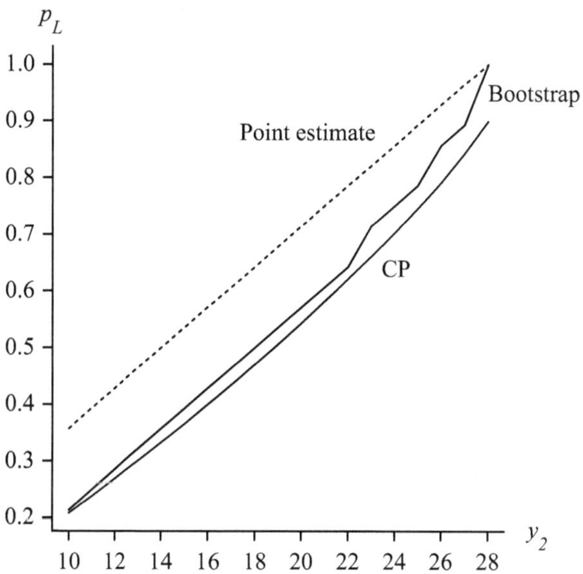

Fig. 15.3. Point estimate (*dashed*), CP lower 95 % confidence bound, and bootstrap lower 95 % confidence bound for the reliability of component 2 based on a sample of size $n_2 = 28$

the system components had one or more failures during their life test. When component i has perfect test results (e.g., $y_i = n_i$), the sample size becomes irrelevant. Thus, a test where two components out of two pass the test is equivalent to one where 100 components out of 100 pass the test from the perspective of the bootstrapping algorithm. This is clearly unacceptable. The next section gives a modification to the bootstrapping approach that adjusts for these perfect tests.

15.4 Perfect Component Test Results

The problem created by perfect component test results is likely to occur for components and systems with moderate to high reliability. As suggested by Chick [30] and Martz and Waller [110, pages 265–266], a beta(α_1, α_2) prior distribution can be placed on the component reliability. The beta distribution is a logical choice for a prior distribution of the component reliability due to (1) the flexibility in the shape of its PDF, (2) its (0, 1) support, and (3) its analytically tractable conjugate posterior distribution. Determining the values of the parameters α_1 and α_2 is a problem that will be addressed in the following two subsections.

The beta distribution has PDF

$$f(x) = \frac{\Gamma(\alpha_1 + \alpha_2)}{\Gamma(\alpha_1)\Gamma(\alpha_2)} x^{\alpha_1 - 1}(1 - x)^{\alpha_2 - 1} \qquad 0 < x < 1,$$

where α_1 and α_2 are positive shape parameters. This is the standard parameterization, although Martz and Waller [110] use a slightly different form. The mean of a beta(α_1, α_2) random variable is

$$\mu = E[X] = \frac{\alpha_1}{\alpha_1 + \alpha_2}$$

and the variance is

$$\sigma^2 = V[X] = \frac{\alpha_1 \alpha_2}{(\alpha_1 + \alpha_2)^2(\alpha_1 + \alpha_2 + 1)}.$$

If the prior distribution of a reliability $P \sim$ beta$(\alpha_1,\ \alpha_2)$ and the sampling is binomial (as it is in our case), then the posterior distribution of P is beta$(\alpha_1 + y,\ \alpha_2 + n - y)$, where n is the number of components on test and y is the number of passes. The difficulty in our case is in determining the appropriate values for α_1 and α_2. For the time being, we will proceed as if we know the values of α_1 and α_2 and give an algorithm for finding the lower reliability confidence bound p_L. Estimating α_1 and α_2 will be addressed subsequently.

A lower reliability confidence bound p_L can be determined by generating a bootstrap beta random variate (rather than a binomial) when the component test results are perfect. An algorithm for determining p_L for a k-component

Table 15.3. Bootstrap algorithm for calculating a $(1-\alpha)100\,\%$ lower confidence bound for the reliability of a k-component series system when some components have perfect test results

```
for j from 1 to B                                  [resampling loop]
  for i from 1 to k                       [loop through components]
    if (y_i = n_i) p~_i <- Beta(α_1i + y_i, α_2i)   [comp. i reliability: perfect test]
    else p~_i <- Binomial(n_i, y_i/n_i)/n_i  [comp. i reliability: failure(s) occur]
  z_j <- ∏_{i=1}^{k} p~_i                  [calculate system reliability]
sort z                                [sort the system reliability values]
return z_αB                            [return estimate for lower bound]
```

series system using B resamplings with some or all components having perfect tests is given Table 15.3. If component i has perfect test results (e.g., $y_i = n_i$) then the analyst must define the prior beta parameters α_{1i} and α_{2i}.

Example 15.2. Next is an example of a three-component series system. Table 15.4 is identical to Table 15.1, except that component 2 now has perfect (28/28) test results. The point estimate for the system reliability increases to

$$\frac{21}{23} \cdot \frac{28}{28} \cdot \frac{82}{84} = \frac{41}{46} \cong 0.8913.$$

Thus, the effect of the one additional component that passed the test increases the system reliability estimate from approximately 0.86 to approximately 0.89. This increase should be reflected in an appropriate increase in the lower confidence limit p_L.

Table 15.4. Failure data for a three-component series system

Component number	$i=1$	$i=2$	$i=3$
Number passing (y_i)	21	28	82
Number on test (n_i)	23	28	84

This algorithm has been implemented as the S-Plus function `seriessystembayesboot` given in Appendix 4 (the number of bootstrap replications $B = 10{,}000$ and values of the beta parameters $\alpha_{12} = 1$ and $\alpha_{22} = 1$ are arbitrary). As before, `n` and `y` are vectors of length k, and `alpha` is a real number between 0 and 1, e.g.,

```
> seriessystembayesboot(c(23, 28, 84), c(21, 28, 82),
                         0.05)
```

prints a point estimate and a 95 % lower confidence interval bound on the system reliability for the three-component series system. After a call to `set.seed(3)` to set the random number seed, five calls to `seriessystembayesboot` yield the following values for p_L:

$$0.7474437 \qquad 0.7484738 \qquad 0.7492014 \qquad 0.7484301$$
$$0.7495972.$$

With the choice $\alpha_1 = \alpha_2 = 1$, the increase of approximately 0.03 in the point estimate of the system reliability from the previous example results in only a tiny increase in the lower confidence interval limits. This is clearly unacceptable. What happened? The arbitrary choice of $\alpha_1 = 1$ and $\alpha_2 = 1$ has resulted in a uniform prior distribution, which is an overly pessimistic assessment of the reliability of component 2, particularly in light of the perfect test results.

What choice would make more sense? It is important to skew the PDF of the beta prior distribution so that its mean is greater than 1/2, or, equivalently, choose $\alpha_2 < \alpha_1$. There are four different shapes of the PDF associated with the choice of parameters that satisfy the constraint $\alpha_2 < \alpha_1$. Most important is the value of the PDF near $f(1)$, since these are the particular reliability values of interest. The following four cases demark various features of the PDF.

- $f(0) = 0$ and $f(1) = 0$ when $1 < \alpha_2 < \alpha_1$ (Case I).
- $f(1)$ is finite when $1 = \alpha_2 < \alpha_1$ (Case II).
- a vertical asymptote at $x = 1$ and $f(0) > 0$ when $\alpha_2 < 1 = \alpha_1$ (Case III).
- a vertical asymptote at $x = 1$ and $f(0) = 0$ when $\alpha_2 < 1 < \alpha_1$ (Case IV).

We have disregarded the case $\alpha_2 < \alpha_1 < 1$ because this results in a vertical asymptote at both 0 and 1, which is inconsistent with the PDF of a high reliability component. The most intuitively appealing of the four cases listed above it the fourth case, $\alpha_2 < 1 < \alpha_1$, since this minimizes the probability of generating a small beta variate (since $f(0) = 0$) and pushes as much of the probability near 1 as possible due to the vertical asymptote near 1.

Table 15.5 gives means of the beta prior distribution and lower confidence interval bounds for several combinations of α_1 and α_2 satisfying the constraint $\alpha_2 < \alpha_1$. The lower bounds are determined by taking the sample median of five runs of `seriessystembayesboot` with $B = 10{,}000$ resampled series systems per run. The subscript on the lower bound indicates which of the shapes in the list given above is represented. The value of the lower bound is quite sensitive to the choices of α_1 and α_2. There are many (α_1, α_2) pairs that yield a reasonable lower bound.

The following two subsections outline methods for estimating the parameters of the prior distribution. The first subsection considers the case when previous test results exist, so data are available to estimate $\hat{\alpha}_1$ and $\hat{\alpha}_2$. The second subsection considers the case when no previous test data are available.

Previous Test Data Exists. When previous test data that is representative of the current test data for a perfect component exists, this data can be fit to yield parameter estimates $\hat{\alpha}_1$ and $\hat{\alpha}_2$ for the prior beta distribution. Let $z_1, z_2, \ldots, z_n$ denote the fraction surviving for previous tests on a componentof interest with equal sample sizes (which has perfect test results and need a beta prior distribution). The maximum likelihood estimators satisfy the simultaneous equations (Evans et al. [50, page 41]):

Table 15.5. Prior beta distribution mean and lower 95 % confidence interval limit estimate for the system reliability

α_1 \ α_2	0.1	1	10
1	$0.909/0.779_{\mathrm{III}}$	—	—
10	$0.990/0.783_{\mathrm{IV}}$	$0.909/0.759_{\mathrm{II}}$	—
100	$0.999/0.783_{\mathrm{IV}}$	$0.990/0.779_{\mathrm{II}}$	$0.909/0.725_{\mathrm{I}}$

$$\psi(\hat{\alpha}_1) - \psi(\hat{\alpha}_1 + \hat{\alpha}_2) = \frac{1}{n}\sum_{i=1}^{n} \log z_i,$$

$$\psi(\hat{\alpha}_2) - \psi(\hat{\alpha}_1 + \hat{\alpha}_2) = \frac{1}{n}\sum_{i=1}^{n} \log(1 - z_i),$$

where ψ is the digamma function. Law and Kelton [89] outline methods for calculating $\hat{\alpha}_1$ and $\hat{\alpha}_2$. These equations have no closed-form solution, and must be solved iteratively. Alternatively, the method of moments estimates are found by equating the population mean μ and population variance σ^2 to the associated sample moments:

$$\bar{z} = \frac{1}{n}\sum_{i=1}^{n} z_i, \qquad s^2 = \frac{1}{n}\sum_{i=1}^{n} (z_i - \bar{z})^2,$$

which results in the closed-form method of moments estimators:

$$\hat{\alpha}_1 = \frac{(1-\bar{z})\bar{z}^2}{s^2} - \bar{z}, \qquad \hat{\alpha}_2 = \frac{\hat{\alpha}_1(1-\bar{z})}{\bar{z}}.$$

Example 15.3. Next is an example of estimating the beta parameters from previous experiments. Consider the previous example, where previous test results on component 2 have yielded the following $n = 4$ fractions surviving:

$$z_1 = \frac{27}{28}, \qquad z_2 = \frac{28}{28}, \qquad z_3 = \frac{26}{28}, \qquad z_4 = \frac{27}{28}.$$

Because the sample mean and variance are

$$\bar{z} = \frac{1}{n}\sum_{i=1}^{n} z_i = \frac{27}{28} \cong 0.964 \text{ and } s^2 = \frac{1}{n}\sum_{i=1}^{n}(z_i - \bar{z})^2 = \frac{1}{1568} \cong 0.000638,$$

the method of moments estimators are (these correspond to Case I from the previous list):

$$\hat{\alpha}_1 = \frac{1431}{28} \cong 51.11, \qquad \hat{\alpha}_2 = \frac{53}{28} \cong 1.89.$$

When these values for the parameters are used in `seriessystembayesboot`, the median of five lower 95 % confidence bounds with $B = 10,000$ for the system reliability is 0.763.

No Previous Test Data Exists. We now turn to the more difficult case of determining the prior beta distribution parameter estimates $\hat{\alpha}_1$ and $\hat{\alpha}_2$ in the case of a component with perfect test results and when no previous test data are available. For such a component, the point estimate of the component reliability is $\hat{p} = 1$ and the CP lower reliability bound is $p_L = \alpha^{1/n}$. One heuristic technique for determining the parameters is to choose $\hat{\alpha}_1$ and $\hat{\alpha}_2$ such that $F(p_L) = \alpha$, i.e.,

$$\int_0^{p_L} \frac{\Gamma(\hat{\alpha}_1 + \hat{\alpha}_2)}{\Gamma(\hat{\alpha}_1)\Gamma(\hat{\alpha}_2)} x^{\hat{\alpha}_1 - 1}(1 - x)^{\hat{\alpha}_2 - 1} dx = \alpha. \tag{15.1}$$

The intuition behind this choice is that $100\alpha\%$ of the time, a prior component reliability (which will be modified subsequently by the data set) will assume a value less than p_L. One problem with this criteria is that there are an infinite number of $\hat{\alpha}_1$ and $\hat{\alpha}_2$ that satisfy this equation. Further refinement is necessary.

For a sample of size $n > 1$, the $(\hat{\alpha}_1, \hat{\alpha}_2)$ pair satisfying Eq. (15.1) will (a) intersect the line $\hat{\alpha}_1 = 1$ on $0 < \hat{\alpha}_2 < 1$ and (b) intersect the line $\hat{\alpha}_2 = 1$ on $\hat{\alpha}_1 > 1$. One technique for determining a $(\hat{\alpha}_1, \hat{\alpha}_2)$ pair is to find the intersection of the values of $\hat{\alpha}_1$ and $\hat{\alpha}_2$ that satisfy Eq. (15.1) and the lines $\hat{\alpha}_1 = 1$ and $\hat{\alpha}_2 = 1$. These two points of intersection, or any point on the line segment connecting them can be used as prior beta distribution parameter estimates. It is interesting to note that

- the intersection of Eq. (15.1) and the line $\hat{\alpha}_1 = 1$ corresponds to Case III for the beta distribution parameters (Scenario 1)
- the intersection of Eq. (15.1) and the line $\hat{\alpha}_2 = 1$ corresponds to Case II for the beta distribution parameters (Scenario 2)

- any point on the line segment connecting the two intersection points (not including the endpoints of the segment) corresponds to Case IV for the beta distribution parameters (Scenario 3)

We first consider the intersection of Eq. (15.1) and $\alpha_1 = 1$. Integration of the beta PDF is analytic in this case, yielding

$$1 - (1 - p_L)^{\hat{\alpha}_2} = \alpha$$

or

$$\hat{\alpha}_2 = \frac{\log(1-\alpha)}{\log(1-\alpha^{1/n})}.$$

Next, we first consider the intersection of Eq. (15.1) and $\alpha_2 = 1$. The integration of the beta PDF is analytic in this case as well, yielding

$$p_L^{\hat{\alpha}_1} = \alpha$$

or

$$\hat{\alpha}_1 = n.$$

Example 15.4. Next is an example of a three-component series system with beta prior distributions. Consider again the three-component series systems. System 1 has test results displayed in Table 15.1. System 2 has test results displayed in Table 15.4. The point estimate for the system reliability of System 1 is

$$\frac{21}{23} \cdot \frac{27}{28} \cdot \frac{82}{84} = \frac{1107}{1288} \cong 0.8595$$

and the point estimate for the system reliability of System 2 is

$$\frac{21}{23} \cdot \frac{28}{28} \cdot \frac{82}{84} = \frac{41}{46} \cong 0.8913.$$

Hence the slight difference between the two test results (the perfect test results for component 2) has resulted in a $0.8913 - 0.8595 = 0.0318$ increase in the point estimate for the system reliability. A similar increase in the lower bound for the system reliability for a reasonable procedure is expected.

The earlier analysis of System 1 using APPL has resulted in an exact (no resampling variability) bootstrap 95 % lower limit on the system reliability of 0.746. Table 15.6 contains 95 % lower confidence limits for the system reliability using four different combinations of prior beta distribution parameter estimates $\hat{\alpha}_1$ and $\hat{\alpha}_2$ for component 2. The parameter estimates for Scenario 3 are found by averaging the parameter estimates for Scenarios 1 and 2. The lower bounds p_L are determined by taking the median result of five runs with $B = 10,000$ replications using the bootstrap procedure described earlier. The column labeled Δp_L gives the difference between the lower confidence limit for System 2 and the lower confidence limit for System 1. The uniform prior model is too wide because the $0.748 - 0.746 = 0.002$ increase in the lower bound is inconsistent with the 0.0318 increase in the point estimate for the system reliability. Based on this example only, Scenarios 1 and 3 seem to be the most appropriate since their increases in the lower bound bracket the increase in the point estimator for the system reliability.

Table 15.6. Lower reliability bounds ($\alpha = 0.05$) for the system reliability of a three-component series system with alternative beta prior parameters

Model	$\hat{\alpha}_1$	$\hat{\alpha}_2$	p_L	Δp_L
Uniform prior	1	1	0.748	0.002
Scenario 1	1	0.022418	0.785	0.039
Scenario 2	28	1	0.769	0.023
Scenario 3	14.5	0.511209	0.772	0.026

Our heuristic, which chooses $\hat{\alpha}_1$ and $\hat{\alpha}_2$ such that $F(p_L) = \alpha$ works reasonably well in the example with one component having perfect test results, but will likely need to be modified if several components have perfect test results. A large-scale Monte Carlo simulation which involves varying α, the number of system components, the configuration of the system components, and the expected fraction of cases where perfect test results are encountered is the only way to evaluate the techniques presented here, and to compare them, for example, with the asymptotic techniques presented in Mann et al. [104, page 498]. Such a simulation is appropriate on a system-by-system basis.

15.5 Simulation

Monte Carlo simulation is used to test several heuristic methods along with the techniques developed in this paper. We begin with a pilot simulation that is used to evaluate a large number of methods in order to thin the number of methods considered.

Simulation Study A. The system considered in this pilot study is a three-component series system with identical components. In keeping with the earlier example, there are $n_1 = 23$, $n_2 = 28$, and $n_3 = 84$ components of each type placed on test. There are $B = 1000$ bootstrap replications used and 1000 simulation replications conducted. The stated coverage of the lower confidence interval bound for the system reliability is 0.95. If the intervals cover approximately 95 % of the true system reliability values for a wide range of true component reliabilities, then the confidence interval procedure is performing adequately. Using the two-sided CP confidence interval procedure for a single component described earlier, the acceptable range for the fraction of simulated confidence intervals (at $\alpha = 0.01$) covering the true system reliability is from 0.931 to 0.968 inclusive. The simulations are run in S-Plus using the `set.seed` command prior to each run to exploit the common random numbers variance reduction technique.

In addition, the number of components that achieve perfect test results is computed for general true component reliabilities p_1, p_2, and p_3. For general n_1, n_2, and n_3, let the random variable W be the number of components with perfect test results. The PDF of W is

$$f(w) = \begin{cases} (1 - p_1^{n_1})(1 - p_2^{n_2})(1 - p_3^{n_3}) & w = 0 \\ (1 - p_1^{n_1})(1 - p_2^{n_2})p_3^{n_3} + (1 - p_1^{n_1})p_2^{n_2}(1 - p_3^{n_3}) & \\ \quad + p_1^{n_1}(1 - p_2^{n_2})(1 - p_3^{n_3}) & w = 1 \\ (1 - p_1^{n_1})p_2^{n_2}p_3^{n_3} + p_1^{n_1}(1 - p_2^{n_2})p_3^{n_3} + p_1^{n_1}p_2^{n_2}(1 - p_3^{n_3}) & w = 2 \\ p_1^{n_1}p_2^{n_2}p_3^{n_3} & w = 3. \end{cases}$$

These values are computed and given in Table 15.7 for various true, identical component reliabilities ranging from 0.60 to 0.99.

Nine algorithms for handling the case of one or more components with perfect test results are compared in the pilot simulation. We have included algorithms of an ad hoc nature (e.g., Algorithms 2 and 3) and those with some theoretical basis (e.g., Algorithm 9) in order to show that the beta prior approach dominates the other approaches as component reliability increases.

- **Algorithm 1:** Pure bootstrapping. A component with a perfect test always generates perfect simulated results.
- **Algorithm 2:** Always assume a failure. When component i has perfect test results (i.e., $y_i = n_i$), introduce an artificial failure by assuming that $y_i = n_i - 1$, for $i = 1, 2, \ldots, n$.
- **Algorithm 3:** Increase the sample size. For perfect test results, artificially increase sample size to approximate the lower confidence bounds with a single failure using the confidence intervals for a single component given earlier in the paper, then bootstrap. In our case, $n_1 = 37$ ($y_1 = 36$), $n_2 = 45$ ($y_2 = 44$), and $n_3 = 134$ ($y_3 = 133$).
- **Algorithm 4:** Bayes bootstrapping with $\alpha_1 = 1$ and $\alpha_2 = 1$ (i.e., uniform prior).
- **Algorithm 5:** Bayes bootstrapping with $\alpha_1 = 1$ for all components, and $\alpha_2 = \log(1 - \alpha)/\log(1 - \alpha^{1/n_i})$, for $i = 1, 2, \ldots, k$, as described earlier.
- **Algorithm 6:** Bayes bootstrapping with $\alpha_1 = n_i$, for $i = 1, 2, \ldots, k$, and $\alpha_2 = 1$ for all components, as described earlier.
- **Algorithm 7:** Bayes bootstrapping with α_1 and α_2 that are averages of the values given in Algorithms 5 and 6.
- **Algorithm 8:** Bayes bootstrapping with $\alpha_1 = 100$ and $\alpha_2 = 1$.
- **Algorithm 9:** A procedure from Mann et al. [104, pages 497–499] which, using asymptotic normal theory, calculates a lower bound as:

$$\prod_{i=1}^{k} \frac{y_i}{n_i} - z_\alpha \sqrt{\prod_{i=1}^{k} \left(\frac{y_i}{n_i}\right)^2 \sum_{j=1}^{k} \left(\frac{1}{y_j} - \frac{1}{n_j}\right)}.$$

The performance of the confidence intervals given in Table 15.7 is as expected. Algorithm 1, for example, which takes the overly optimistic, pure bootstrapping approach, produces lower confidence limits that are shifted up, resulting in fewer-than-expected lower confidence limits that fall below the true system reliability. The opposite case is true for the pessimistic uniform prior distribution in Algorithm 4. In fact, once the Bayesian portion of the algorithm began to dominate (i.e., when the component reliabilities are large), all nine algorithms fail to deliver confidence intervals with the appropriate coverage. We experimented with the confidence interval that performed the best (Algorithm 7, which averages the parameter estimates of Algorithms 5 and 6) by replacing "average" with "linear combination," but did not produce results that were significantly better than those presented in Table 15.7.

The abysmal performance of all of these algorithms for high-reliability components is consistent with the work of Martz and Duran [109], who considered lower confidence bounds for the system reliability of 20 system configurations and component reliabilities using three algorithms and two values of α (0.05 and 0.10). Their intervals also diverged from the stated coverages.

Simulation Study B. This poor performance led us to re-code our algorithms in C and to do an exhaustive search in the (α_1, α_2) plane for values of the beta prior parameters α_1 and α_2 that yield reasonable coverages for lower confidence bounds on the system reliability. We returned to the case of a single component. Figure 15.4 shows the results of this exhaustive search for $n = 23$ components on test. Every (α_1, α_2) pair that resulted in a confidence interval whose coverage did not statistically differ from 0.95 was plotted for $p = 0.91, 0.92, \ldots, 0.99$. For each particular population reliability p shown in Figure 15.4, the areas where appropriate coverages are achieved are quite narrow. Unfortunately, the graph in Figure 15.4 shows that there is no single (α_1, α_2) pair that will work for all values of p.

The following procedure has been developed as a compromise that allows reasonable lower confidence limit coverage in the case of a system with one or more components having perfect test results:

- For each of the components in the system, consult with someone familiar with the component to get a point estimate of the component reliability p_i^*, $i = 1, 2, \ldots, k$.
- Determine the number of components to be tested $n_1, n_2, \ldots, n_k$.
- For each (p_i^*, n_i) pair, perform an exhaustive search of the (α_1, α_2) plane to find a $(\tilde{\alpha}_1, \tilde{\alpha}_2)$ pair which yields appropriate coverage.
- Perform Bayesian bootstrapping as described earlier in this paper using the test results (n_i, y_i) and the appropriate $(\tilde{\alpha}_1, \tilde{\alpha}_2)$ values from the previous step.

Table 15.7. Estimated lower confidence interval coverage ($\alpha = 0.05$) for the system reliability of a three-component series system with 1000 replications using bootstrapping with a beta prior distribution for perfect test results. The random variable W denotes the number of components with perfect test results. The tabled values give the fraction of intervals that fall below the true system reliability. Fractions set in boldface type are in the range 0.931 to 0.968 inclusive and are not statistically different from the stated coverage of 0.95

True reliability:	0.60	0.70	0.80	0.90	0.91	0.92	0.93	0.94	0.95	0.96	0.97	0.98	0.99
$\Pr(W = 0)$	0.999	0.999	0.992	0.864	0.822	0.767	0.704	0.621	0.521	0.401	0.267	0.131	0.029
$\Pr(W = 1)$	10^{-5}	10^{-4}	0.008	0.132	0.170	0.216	0.271	0.334	0.401	0.460	0.483	0.424	0.221
$\Pr(W = 2)$	10^{-12}	10^{-8}	10^{-5}	0.005	0.008	0.014	0.025	0.044	0.078	0.136	0.234	0.380	0.492
$\Pr(W = 3)$	10^{-30}	10^{-21}	10^{-13}	10^{-6}	10^{-6}	10^{-5}	10^{-4}	10^{-4}	0.001	0.004	0.016	0.065	0.257
Algorithm 1	**0.958**	**0.964**	**0.946**	0.912	0.910	0.910	0.899	0.899	0.878	0.873	0.809	0.765	0.738
Algorithm 2	**0.958**	**0.964**	**0.945**	**0.950**	**0.960**	0.980	0.988	1.000	1.000	1.000	1.000	1.000	1.000
Algorithm 3	**0.958**	**0.964**	**0.953**	**0.939**	**0.954**	**0.937**	**0.955**	0.972	0.981	1.000	1.000	1.000	1.000
Algorithm 4	**0.958**	**0.964**	**0.948**	**0.960**	**0.957**	**0.968**	0.988	1.000	1.000	1.000	1.000	1.000	1.000
Algorithm 5	**0.958**	**0.964**	**0.939**	0.927	0.918	0.919	0.902	0.910	0.895	0.877	0.832	0.820	0.745
Algorithm 6	**0.958**	**0.964**	**0.947**	0.923	**0.942**	**0.951**	**0.947**	**0.947**	0.970	0.976	1.000	1.000	1.000
Algorithm 7	**0.958**	**0.964**	**0.933**	0.923	**0.947**	0.917	**0.944**	**0.942**	**0.936**	**0.936**	0.982	1.000	1.000
Algorithm 8	**0.958**	**0.964**	**0.951**	0.925	0.925	0.916	0.911	0.926	0.900	0.906	0.876	0.827	1.000
Algorithm 9	0.975	**0.947**	**0.949**	0.920	0.918	0.919	0.888	0.897	0.886	0.867	0.822	0.737	0.730

Example 15.5. The final example illustrates this technique for a single-component system. Figure 15.5 contains a graph of the actual coverage for 1000 simulation replications of a single-component system with $n = 28$ and a three-component system with $n_1 = 23$, $n_2 = 28$, and $n_3 = 84$.

All true component reliabilities are equal, and plotted on the horizontal axis. The stated coverage on all intervals is 0.95. The usual bounds around 0.95 (at 0.931 and 0.968) which denote confidence intervals whose actual coverage does not differ significantly from the specification are given as horizontal dashed lines. All Bayesian procedures use $(\alpha_1, \alpha_2) = (252.28, 4.67)$, which were values that fell outside of the axes in Figure 15.4 associated with the $p^* = 0.97$ estimate for the reliability of the second component. The jagged appearance for the coverage for the interval for a single component (dotted) is consistent with the same pattern shown by Blyth [17]. The three-component system (solid), on the other hand, has different numbers of components on test that seem to "average out" these fluctuations, resulting in appropriate coverage through $p = 0.97$.

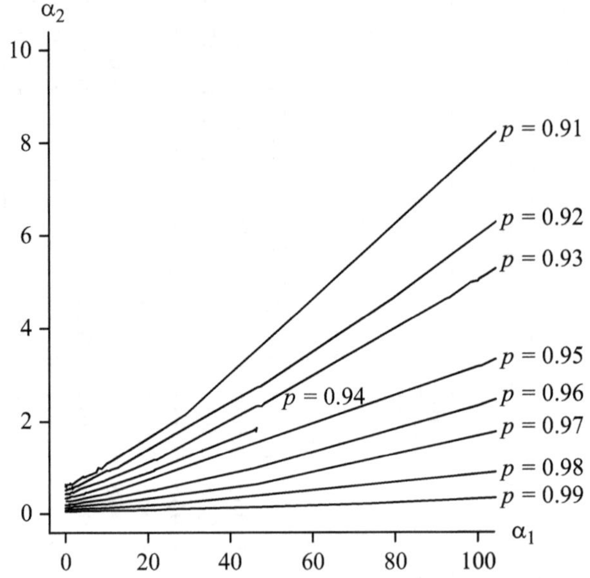

Fig. 15.4. Prior distribution parameter pairs that give accurate coverage for the example as a function of the reliability p when $n = 23$

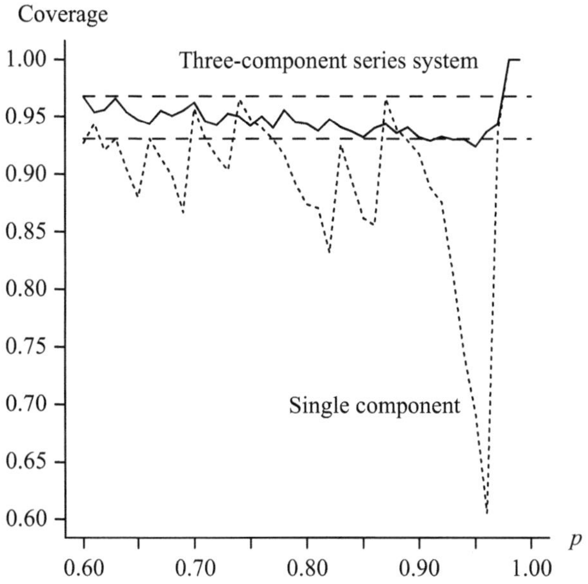

Fig. 15.5. Lower 95 % confidence bound coverage for a single component system (*dotted*) and a three-component system (*solid*) and region not statistically different from the specification (*dashed*)

15.6 Conclusions

Determining lower confidence bounds from binary data remains an important yet elusive question. The Bayesian bootstrapping procedures developed here yield adequate coverages given that an expert is able make a good initial estimate of the reliabilities of individual components. The estimates discussed here improve with increasing system complexity.

Acknowledgments

The author gratefully acknowledges the helpful interactions with John Backes, Don Campbell, Mike Crawford, Mark Eaton, Jerry Ellis, Greg Gruver, Jim Scott, Bob Shumaker, Bill Treadwell, and Mark Vinson concerning the paper and helpful comments from two anonymous referees and the editor.

Appendix 1

S-Plus code for calculating a CP $(1-\alpha)100\,\%$ lower confidence bound for a single component for n components on test and y passes. This function was used to generate the lower confidence bounds in Example 15.1.

```
> confintlower <- function(n, y, alpha) {
>   if (y == 0) {
>     pl <- 0
>   }
>   if (y == n) {
>     pl <- alpha ^ (1 / n)
>   }
>   if (y > 0 && y < n) {
>     fcrit1 <- qf(alpha, 2 * y, 2 * (n - y + 1))
>     pl <- 1 / (1 + (n - y + 1) / (y * fcrit1))
>   }
>   pl
> }
```

Appendix 2

S-Plus code for calculating a bootstrap $(1-\alpha)100\,\%$ lower confidence interval bound for a k-component series system of independent components using B bootstrap replications. This implements the algorithm given in Table 15.2.

```
> seriessystemboot <- function(n, y, alpha) {
>   k <- length(n)
>   b <- 10000
>   z <- rep(1, b)
>
>   point <- prod(y) / prod(n)
>
>   for (j in 1:b) {
>     for (i in 1:k) {
>      z[j] <- z[j] * rbinom(1, n[i], y[i] / n[i]) / n[i]
>     }
>   }
>   z <- sort(z)
>   pl <- z[floor(alpha * b)]
>   c(point, pl)
> }
```

Appendix 3

APPL code for calculating a bootstrap $(1-\alpha)100\,\%$ lower confidence interval bound for a k-component series system of independent components using the equivalent of $B = +\infty$ bootstrap replications.

```
> n1 := 23;
> y1 := 21;
> X1 := BinomialRV(n1, y1 / n1);
> X1 := Transform(X1, [[x -> x / n1], [-infinity, infinity]]);
> n2 := 28;
> y2 := 27;
> X2 := BinomialRV(n2, y2 / n2);
> X2 := Transform(X2, [[x -> x / n2], [-infinity, infinity]]);
> n3 := 84;
> y3 := 82;
> X3 := BinomialRV(n3, y3 / n3);
> X3 := Transform(X3, [[x -> x / n3], [-infinity, infinity]]);
> Temp := Product(X1, X2);
> T := Product(Temp, X3);
```

Appendix 4

S-Plus code for calculating a bootstrap $(1-\alpha)100\,\%$ lower confidence interval bound for a k-component series system of independent component with some perfect component test results using B bootstrap replications. This implements the algorithm given in Table 15.3.

```
> seriessystembayesboot <- function(n, y, alpha) {
>   k <- length(n)
>   alpha1 <- 1
>   alpha2 <- 1
>   b <- 10000
>   z <- rep(1, b)
>
>   point <- prod(y) / prod(n)
>
>   for (j in 1:b) {
>     for (i in 1:k) {
>       if (y[i] == n[i]) z[j] <- z[j] * rbeta(1, alpha1 + y[i],
>           alpha2)
>       else z[j] <- z[j] * rbinom(1, n[i], y[i] / n[i]) / n[i]
>     }
>   }
>   z <- sort(z)
>   pl <- z[floor(alpha * b)]
>   c(point, pl)
> }
```

References

1. Aalen, O. O., & Gjessing, H. K. (2001). Understanding the shape of the hazard rate: A process point of view. *Statistical Sciences*, *16*(1), 1–22.
2. Agresti, A., & Coull, B. (1998). Approximate is better than 'exact' for interval estimation of binomial proportions. *The American Statistician*, *52*(395), 119–126.
3. Anderson, K., & Rønn, B. (1995). A nonparametric test for comparing two samples where all observations are either left- or right-censored. *Biometrics*, *51*(1), 323–329.
4. Arnold, H. J. (1965). Small sample power of the one sample Wilcoxon test for non-normal shift alternatives. *The Annals of Mathematical Statistics*, *36*(6), 1767–1778.
5. Bae, S. J., Kuo, W., & Kvam, P. H. (2007). Degradation models and implied lifetime distributions. *Reliability Engineering and System Safety*, *92*, 601–608.
6. Balasooriya, U. (1995). Failure-censored reliability sampling plans for the exponential distribution. *Journal of Statistical Computation and Simulation*, *52*, 337–349.
7. Barlow, R. E., & Proschan, F. (1981). *Statistical theory of reliability and life testing*. Silver Spring: To Begin With.
8. Barr, D. R., & Davidson, T. (1973). A Kolmogorov–Smirnov test for censored samples. *Technometrics*, *15*, 739–757.
9. Barr, D. R., & Zehna, P. (1983). *Probability: Modeling uncertainty*. Reading: Addison-Wesley.
10. Basawa, I. V., & Rao, B. L. S. P. (1980). *Statistical inference for stochastic processes*. London: Academic.
11. Bellera, C. A., Julien, M., & Hanley, J. A. (2010). 'Normal approximations to the distribution of the Wilcoxon statistics: Accurate to what N? graphical insights. *Journal of Statistics Education*, *18*(2), 1–17.

A.G. Glen, L.M. Leemis (eds.), *Computational Probability Applications*,
International Series in Operations Research & Management Science 247,
DOI 10.1007/978-3-319-43317-2

12. Bergmann, R., Ludbrook, J., & Spooren, W. P. J. M. (2000). Different outcomes of the Wilcoxon–Mann–Whitney test from different statistics packages. *The American Statistician*, *54*(1), 72–77.
13. Billingsley, P. (1995). *Probability and measure*. New York: Wiley.
14. Biswas, A., & Sundaram, R. (2006). Kernel survival function estimation based on doubly censored data. *Communications in Statistics: Theory and Methods*, *35*, 1293–1307.
15. Blair, R. C., & Higgins, J. J. (1985). Comparisons of the power of the paired samples t test to that of Wilcoxon's signed-ranks test under various population shapes. *Psychological Bulletin*, *97*(1), 119–128.
16. Block, A. D., & Leemis, L. M. (2008). Model discrimination for heavily censored survival data. *IEEE Transactions on Reliability*, *57*, 248–259.
17. Blyth, C. R. (1986). Approximate binomial confidence limits. *Journal of the American Statistical Association*, *81*(395), 843–855.
18. Boag, J. W. (1949). Maximum likelihood estimates of the proportion of patients cured by cancer therapy. *Journal of the Royal Statistical Society*, *11*, 15–53.
19. Bousquet, N., Bertholon, H., & Celeux, G. (2006). An alternative competing risk model to the Weibull distribution for modeling aging in lifetime data analysis. *Lifetime Data Analysis*, *12*, 481–504.
20. Bowman, A. W., & Azzalini, A. (1997). *Applied smoothing techniques for data analysis*. Oxford: Oxford University Press.
21. Bridge, P. D., & Sawilowsky, S. S. (1999). Increasing Physicians' awareness of the impact of statistics on research outcomes: Comparative power of the t test and Wilcoxon rank-sum test in small samples applied research. *Journal of Clinical Epidemiology*, *52*(3), 229–235.
22. Büning, H., & Qari, S. (2006). Power of one-sample location tests under distributions with equal Lévy distance. *Communications in Statistics: Simulation and Computation*, *35*(3), 531–545.
23. Burr, I. W. (1967). The effect of non-normality on constants for $\bar{\mathrm{X}}$ and R charts. *Industrial Quality Control*, *23*(11), 563–569.
24. Burr, I. W. (1968). On a general system of distributions: The sample range. *Journal of the American Statistical Association*, *63*(322), 636–643.
25. Burr, I. W. (1976). *Statistical quality control methods*. New York: Marcel Dekker.
26. Brutman, L. (1998). Lebesgue functions for polynomial interpolation: A survey. *Annals of Numerical Mathematics, 4*, 111–127.
27. Caroni, C. (2002). The correct 'ball bearings' data. *Lifetime Data Analysis*, *8*, 395–399.
28. Casella, G., & Berger, R. (2002). *Statistical inference* (2nd ed.). Belmont: Duxbury.
29. Chiang, C. L. (1968). *Introduction to stochastic processes in biostatistics*. New York: Wiley.
30. Chick, S. (2001). personal communication.

31. Cheng, R. C. H. (2006). Resampling methods. In S. Henderson, & B. L. Nelson (Eds.), *Handbooks in operations research and management science: Simulation* (pp. 415–453). Amsterdam: Elsevier.
32. Cho, S. K., & Spiegelberg-Planer, R. (2002). Country nuclear power profiles. http, //www-pub.iaea.org/MTCD/publications/PDF/cnpp2003/CNPP_Webpage/PDF/2002/index.htm. Accessed 6 Dec 2007.
33. Conover, W. J., & Iman, R. L. (1981). Rank transformations as a bridge between parametric and nonparametric statistics. *The American Statistician, 35*(3), 124–129.
34. Cook, T. M., & Russell, R. A. (1989). *Introduction to management science* (4th ed.). Englewood Cliffs: Prentice Hall.
35. Cox, D. R., & Oakes, D. (1984). *Analysis of survival data*. London: Chapman & Hall/CRC.
36. Craig, C. C. (1936). A new exposition and chart for the pearson system of frequency curves. *Annals of Mathematical Statistics, 7*, 16–28.
37. Crowder, M. J. (2001). *Classical competing risks*. London: Chapman & Hall/CRC Press.
38. Crowder, M. J., Kimber, A. C., Smith, R. L., & Sweeting, T. J. (1991). *Statistical analysis of reliability data*. London: Chapman & Hall/CRC.
39. Cui, J. (1999). Nonparametric estimation of a delay distribution based on left-censored and right–truncated data. *Biometrics, 55*(2), 345–349.
40. D'Agostino, R. B., & Stephens, M. A. (1986). *Goodness-of-fit techniques*. New York: Marcel Dekker.
41. David, H. A. (1993). A note on order statistics for dependent variables. *The American Statistician, 47*, 198–199.
42. David, H. A., & Moeschberger, M. L. (1978). *The theory of competing risks*. Griffin's statistical monographs and courses (Vol. 39). New York: Macmillan.
43. David, H. A., & Nagaraja, H. N. (2003). *Order statistics* (3rd ed.). Hoboken, NJ: Wiley.
44. Dembo, A., & Zeitouni, O. (1998). *Large deviations techniques and applications*. Heidelberg: Springer.
45. Deshpande, J. V., & Purohit, S. G. (2005). *Life time data: Statistical models and methods*. London: World Scientific.
46. Drew, J.H., Evans, D., Glen, A., & Leemis, L. (2008). *Computational probability: Algorithms and applications in the mathematical sciences*. New York: Springer.
47. Drew, J. H., Glen, A. G., & Leemis, L. M. (2000). Computing the cumulative distribution function of the Kolmogorov–Smirnov statistic. *Computational Statistics and Data Analysis 34*, 1–15.
48. Durbin, J. (1975). Kolmogorov–Smirnov tests when parameters are estimated with applications to tests of exponentiality and tests on spacings. *Biometrika, 62*, 5–22.
49. Efron, B., & Tibshirani, R. J. (1993). *An introduction to the bootstrap*. New York, NY: Chapman & Hall.

50. Evans, G., Hastings, N., & Peacock, B. (2000). *Statistical distributions* (3rd ed.) New York: Wiley.
51. Evans, D. L., & Leemis, L. M. (2000). Input modeling using a computer algebra system. In *Proceedings of the 2000 Winter Simulation Conference* (pp. 577–586).
52. Evans, D., Leemis, L., & Drew, J. (2006). The distribution of order statistics for discrete random variables with applications to bootstrapping. *INFORMS Journal on Computing*, *18*, 19–30.
53. Feller, W. (1957). *An introduction to probability theory and its applications*. New York: Wiley.
54. Feller, W. (1971). *An introduction to probability theory and its applications* (Vol. II). New York: Wiley.
55. Forbes, C., Evans, M., Hastings, N., & Peacock, B. (2011). *Statistical distributions* (4th ed.). Hoboken: Wiley.
56. Gehan, E. A. (1965). A generalized Wilcoxon test for comparing arbitrarily singly-censored samples. *Biometrika*, *52*, Parts 1 and 2, 203–223.
57. Gelman, A., Carling, J., Stern, H., & Rubin, D. (2004). *Bayesian data analysis* (2nd ed.). New York: Chapman & Hall/CRC.
58. Gertsbakh, I. (1995). On the Fisher information in type-i censored and quantal response data. *Statistics and Probability Letters*, *23*(4), 297–306.
59. Glen, A., Barr, D., & Leemis, L. (2001). Order statistics in goodness of fit testing. *IEEE Transactions on Reliability 50*(2), 209–213.
60. Glen, A., Evans, D., & Leemis, L. (2001). APPL: A probability programming language. *The American Statistician*, *55*, 156–166.
61. Glen, A., & Leemis, L. (1997). The arctangent survival distribution. *The Journal of Quality and Technology*, *29*(2), 205–210.
62. Goto, F. (1996). Achieving semiparametric efficiency bounds in left-censored duration models. *Econometrica*, *64*(2), 439–442.
63. Gupta, R., & Kundu, D. (2006). On the comparison of Fisher information of the Weibull and GE distributions. *Journal of Statistical Planning and Inference*, *136*, 3130–3144.
64. Hájek, J., Šidák, Z., & Sen, P. K. (1999). *Theory of rank tests* (2nd ed.). New York: Academic.
65. Heck, A. (1996). *Introduction to maple* (2nd ed.). Berlin: Springer.
66. Hegazy, Y., Green, J. (1975). Some new goodness-of-fit tests using order statistics. *Applied Statistics*, *24*(3), 299–308.
67. Hogg, R. V., McKean, J. W., & Craig, A. T. (2005). *Introduction to the mathematical statistics* (6th ed.). Upper Saddle River, NJ: Prentice–Hall.
68. Hollander, M., & Wolfe, D. A. (1973). *Nonparametric statistical methods.* New York: Wiley.
69. Jiang R., & Jardine, A. K. S. (2006). Composite scale modeling in the presence of censored data. *Reliability Engineering & System Safety*, *91*(7), 756–764.

70. Jiang, S. T., Landers, T. L., & Rhoads, T. R. (2005). Semi-parametric proportional intensity models robustness for right-censored recurrent failure data. *Reliability Engineering & System Safety*, *90*(1), 91–98.
71. Johnson, N. L., Kotz, S., & Balakrishnan, N. (1994). *Continuous univariate distributions* (Vol. 1). New York: Wiley.
72. Johnson, N. L., Kotz, S., & Balakrishnan, N. (1995). *Continuous univariate distributions* (2nd ed., Vol. 2). New York: Wiley.
73. Johnson, R. A., & Wichern, D. W. (1992). *Applied multivariate statistical analysis* (3rd ed.). Upper Saddle River, NJ: Prentice–Hall.
74. Kalbfleisch, J. D., & MacKay, R. J. (1979). On constant-sum models for censored survival data. *Biometrika*, *66*, 87–90.
75. Kalbfleisch, J. D., & Prentice, R. L. (2002). *The statistical analysis of failure time data* (2nd ed.). Hoboken, NJ: Wiley.
76. Kanji, G. K. (2006). *100 statistical tests* (3rd ed.). London: SAGE.
77. Kendall, M. G., & Stuart, A. (1963). *The advanced theory of statistics: Volume II inference and relationship*. New York: Hafner.
78. Kleiber, C., & Kotz, S. (2003). *Statistical size distributions in economics and actuarial sciences*. New York: Wiley.
79. Klein, J., & Moeschberger, M. (1997). *Survival analysis: Techniques for censored and truncated data*. New York: Springer.
80. Klotz, J. (1963). Small sample power and efficiency for the one sample Wilcoxon and normal score tests. *Annals of Mathematical Statistics*, *33*, 498–512.
81. Koch, K. (2007). *Introduction to Bayesian statistics*. Heidelberg: Springer.
82. Kotz, S., & Johnson, N. L. (2006). *Encyclopedia of statistical sciences* (Vol. 5). New York: Wiley.
83. Kuiper, N. H. (1960). Tests concerning random points on a circle. In: *Proc. Koninkljke Nederlandse Akadamie van Wetenschappen, Series A* (Vol. 63, pp. 38–47).
84. Kundu, D., & Sarhan, A. M. (2006). Analysis of incomplete data in presence of competing risks among several groups. *IEEE Transactions on Reliability*, *55*(2), 262–269.
85. Lai, C., & Xie, M. (2006). *Stochastic aging and dependence for reliability*. New York: Springer.
86. Lan, Y., & Leemis, L. (2008). The Logistic–Exponential survival distribution. *Naval Research Logistics*, *55*, 254–264.
87. Larsen, R. J., & Marx, M. L. (2001). *An introduction to mathematical statistics and its applications* (3rd ed.). Upper Saddle River: Prentice–Hall.
88. Law, A. M. (2007). *Simulation modeling and analysis* (4th ed.). New York: McGraw–Hill.
89. Law, A. M., & Kelton, W. D. (2000). *Simulation modeling and analysis* (3rd ed.). New York: McGraw–Hill.

90. Lawless, J. F. (1980). Inference in the generalized gamma and log-gamma distributions. *Technometrics*, *22*(3), 409–419.
91. Lawless, J. F. (2003). *Statistical models and methods for lifetime data* (2nd ed.). New York: Wiley.
92. Lee, E. T. (1992). *Statistical methods for survival data analysis*. New York: Wiley.
93. Leemis, L. (1995). *Reliability: Probabilistic models and statistical methods*. Upper Saddle River: Prentice–Hall.
94. Leemis, L., & Evans, D. (2004). Algorithms for computing the distributions of sums of discrete random variables. *Mathematical and Computer Modeling*, *40*(13), 1429–1452.
95. Leemis, L., & McQueston, J. (2008). Univariate distribution relationships. *The American Statistician*, *62*(1), 45–53.
96. Leemis, L., & Park, S. (2006). *Discrete event simulation: A first course*. Upper Saddle River: Pearson Prentice–Hall.
97. Leemis, L., & Shih, L. (1989). Exponential parameter estimation for data sets containing left- and right-censored observations. *Communications in Statistics – Simulation*, *18*(3), 1077–1085.
98. Leemis, L. M., & Trivedi, K. S. (1996). A comparison of approximate interval estimators for the Bernoulli parameter. *The American Statistician*, *50*(1), 63–68.
99. Lehmann, E. L. (1959). *Testing statistical hypotheses*. New York: Wiley.
100. Li, C. H., & Fard, N. (2007). Optimum bivariate step-stress accelerated life test for censored data. *IEEE Transactions on Reliability*, *56*(1), 77–84.
101. Lieblein, J., & Zelen, M. (1956). Statistical investigation of the fatigue life of deep-groove ball bearings. *Journal of Research of the National Bureau of Standards*, *57*, 273–316.
102. Lilliefors, H. W. (1967). On the Kolmogorov–Smirnov test for normality with mean and variance unknown. *Journal of the American Statistical Association*, *62*, 399–402.
103. Lilliefors, H. W. (1969). On the Kolmogorov–Smirnov test for the exponential distribution with mean unknown. *Journal of the American Statistical Association*, *64*, 387–389.
104. Mann, N. R., Schafer, R. E., & Singpurwalla, N. D. (1974). *Methods for statistical analysis of reliability and life data*. New York: Wiley.
105. Mann, H. B., & Whitney, D. R. (1947). On a test of whether one of two random variables is stochastically larger than the other. *Annals of Mathematical Statistics*, *18*, 50–60.
106. Marsaglia, G., Tsang, W. W., & Wang, J. (2003). Evaluating Kolmogorov's distribution. *Journal of Statistical Software*, *8*(18). http.www.jstatsoft.org/v08/i18/
107. Marshall, A. W., & Olkin, I. (2007). *Life distributions: Structure of nonparametric, semiparametric, and parametric families*. New York: Springer.

108. Martin, M. A. (1990). On bootstrap iteration for coverage correction in confidence intervals. *Journal of the American Statistical Association, 85*(412), 1105–1118.
109. Martz, H. F., & Duran, B. S. (1985). A comparison of three methods for calculating lower confidence limits on system reliability using binomial component data. *IEEE Transactions on Reliability, 34*(2), 113–120.
110. Martz, H. F., & Waller, R. A. (1982). *Bayesian reliability analysis*. New York: Wiley.
111. McKay, A. T., & Pearson, E. S. (1933). A note on the distribution of range in samples of n. *Biometrika, 25*, 415–420.
112. McCornack, R. L. (1965). Extended tables of the Wilcoxon matched pair signed rank statistic. *Journal of the American Statistical Association, 60*, 864–871.
113. Meeker, W., & Escobar, L. (1998). *Statistical methods for reliability data*. New York: Wiley.
114. Michael, J. R., & Schucany, W. R. (1979). A new approach to testing goodness of fit for censored samples. *Technometrics, 21*(4), 435–441.
115. Milevsky, M. A., & Posner, S. E. (1998). Asian options, the sum of log-normals, and the reciprocal gamma distribution. *The Journal of Financial and Quantitative Analysis, 33*(3), 203–218.
116. Mundry, R., & Fischer, J. (1998). Use of statistical programs for nonparametric tests of small samples often leads to incorrect values: Examples from *animal behavior. Animal Behavior, 56*(1), 256–259.
117. Nelson, L. S. (1963). Precedence life test. *Technometrics, 5*, 491–499.
118. Nelson, L. S. (1993). Tests on early failures–the precedence life test. *Journal of Quality Technology, 25*(2), 140–143.
119. Nelson, W. (1982). *Applied life data analysis*. New York: Wiley.
120. Nelson, W. (1986). *How to analyze data with simple plots*. ASQC basic references in quality control: Statistical techniques (Vol. 1) Milwaukee: American Society for Quality Control.
121. Newcombe, R. G. (2001). Logit confidence intervals and the inverse sinh transformation. *The American Statistician, 55*(3), 200–202.
122. Nguyen, N. C., Patera, A. T., & Peraire, J. (2007). A "best points" interpolation method for efficient approximation of parameterized functions. *International Journal for Numerical Methods in Engineering, 73*(4), 521–543.
123. Odell, P., Anderson, K., & D'Agostino, R. (1992). Maximum likelihood estimation for interval-censored data using a Weibull-based accelerated failure time model. *Biometrics, 48*, 951–959.
124. Oller, R., Gomez, G., & Luz Calle, M. (2004). Interval censoring: Model characterizations for the validity of the simplified likelihood. *The Canadian Journal of Statistics, 32*(3), 315–326.
125. O'Reilly, F. J., & Stephens, M. A. (1988). Transforming censored samples for testing fit. *Technometrics, 30*(1), 79–86.

126. Padgett, W. J., & Tomlinson, M. A. (2003). Lower confidence bounds for percentiles of Weibull and Birnbaum–Saunders distributions. *Journal of Statistical Computation and Simulation*, *73*(6), 429–443.
127. Pagano, R. R. (1981). *Understanding statistics in the behavioral sciences*. Saint Paul: West Publishing Co.
128. Park, C. (2005). Parameter estimation of incomplete data in competing risks using the EM algorithm. *IEEE Transactions on Reliability*, *54*(2), 282–290.
129. Parzen, E. (1962). *Stochastic processes*. San Francisco: Holden–Day.
130. Pearson, E. S., & Hartley, H. O. (1951). Moment constants for the distribution of range in normal samples. *Biometrika*, *38*, 463–464.
131. Phillips, M. J., & Sweeting, T. J. (1996). Estimation for censored exponential data when the censoring times are subject to error. *Journal of the Royal Statistical Society, Series B (Methodological)*, *58*(4), 775–783.
132. Pintilie, M. (2006). *Competing risks: A practical perspective.* New York: Wiley.
133. Poirier, D. J. (1995). *Intermediate statistics and econometrics*. Cambridge: MIT Press.
134. Prentice, R. L., & Kalbfleisch, J. D. (1978). The analysis of failure times in the presence of competing risks. *Biometrics*, *34*, 541–554.
135. Rigdon, S., & Basu, A. P. (2000). *Statistical methods for the reliability of repairable systems*. New York: Wiley.
136. Rivlin, T. J. (1981). *An introduction to the approximation of functions*. New York: Dover.
137. Robert, C. (2007). *The Bayesian choice* (2nd ed.). New York: Springer.
138. Rodriguez, R. N. (1977). A guide to the Burr type XII distributions. *Biometrika*, *64*, 129–134.
139. Rose, C., & Smith, M. D. (2002). *Mathematical statistics and mathematica*. Heidelberg: Springer.
140. Ross, S. M. (1985). *Introduction to probability models*. New York: Academic.
141. Ross, S. M. (2006). *Simulation.* New York: Elsevier Academic.
142. Rosenblatt, M. (1952). Remarks on a multivariate transformation. *Annals of Mathematical Statistics*, *23*(3), 470–472.
143. Samson, A., Lavielle, M., & Mentre, F. (2006). Extension of the SAEM algorithm to left-censored data in nonlinear mixed–effects model: Application to HIV dynamics model. *Computational Statistics and Data Analysis*, *51*, 1562–1574.
144. Sarhan, A. M. (2007). Analysis of incomplete, censored data in competing risks models with generalized exponential distributions. *IEEE Transactions on Reliability*, *56*(1), 132–138.
145. Shewhart, W. A. (1931). *Economic control of quality of manufacturing processes*. New York: Van Nostrand.
146. Siegel, S. (1957). Nonparametric statistics. *The American Statistician*, *11*(3), 13–19.

147. Silverman, B. W. (1986). *Density estimation for statistics and data analysis*. London: Chapman and Hall.
148. Smith, S. J. (2006). Lebesgue constants in polynomial interpolation. *Annales Mathimaticae et Informaticae, 3*, 109–123.
149. Soliman, A. A. (2005). Estimation of parameters of life from progressively censored data using Burr-XII model. *IEEE Transactions on Reliability, 54*(1), 34–42.
150. Song, W. T., & Chen, Y. (2011). Eighty univariate distributions and their relationships displayed in a matrix format. *IEEE Transactions on Automatic Control, 56*(8), 1979–1984.
151. Stephens, M. A. (1965). The goodness-of-fit statistics V_N: Distribution and significance points. *Biometrika, 52*, 309–322.
152. Stephens, M. A. (1974). EDF statistics for goodness of fit and some comparisons. *Journal of the American Statistical Association, 69*(347), 730–737.
153. Stephens, M. A. (1986). Tests based on EDF statistics. In R. B. D'Agostino, & M. A. Stephens (Eds.), *Goodness-of-fit techniques* (pp. 97–193), New York: Marcel Dekker.
154. Stewart, J. (1999). *Calculus: Early transcendentals* (4th ed.). Belmont: Brooks/Cole Publishing Company.
155. Streitberg, B., & Rohmel, J. (1986). Exact distributions for permutation and rank tests: An introduction to some recently published algorithms. *Statistical Software Newsletter, 1*, 10–17.
156. Stuart, A., & Ord, K. (1994). *Kendall's advanced theory of statistics* (Vol. 1). London: Hodder Arnold.
157. Sun, J. (2004). Statistical analysis of doubly interval-censored failure time data. *Handbook of statistics* (Vol. 3). Amsterdam: Elsevier.
158. Tadikamalla, P. R. (1980). A look at the Burr and related distributions. *International Statistical Review, 48*, 337–344.
159. Tadikamalla, P. R., Banciu, M., & Popescu, D. (2008). An improved range chart for normal and long-tailed symmetrical distributions. *Naval Research Logistics, 55*, 91–99.
160. Tietjen, G. L. (1986). The analysis and detection of outliers. In R. B. D'Agostino, & M. A. Stephens (Eds.), *Goodness-of-fit techniques* (pp. 497–522). New York: Marcel Dekker.
161. Tippett, L. H. C. (1925). On the extreme individuals and the range of samples taken from a normal population. *Biometrika, 17*, 364–387.
162. U. S. Patent Office (2006). Patent 7010463, Method of determining whether an improved item has a better mean lifetime than an existing item. Inventors Andrew G. Glen and Bobbie L. Foote, assigned to the United States Government.
163. Vargo, E., Pasupathy, R., & Leemis, L. (2010). Moment-ratio diagrams for univariate distributions. *Journal of Quality Technology, 42*(3), 276–286.

164. Vollset, S. E. (1993). Confidence intervals for a binomial proportion. *Statistics in Medicine*, *12*, 809–824.
165. Wheeler, D. J. (2000). *Normality and the process behavior chart*. Knoxville: SPC Press.
166. Wilcoxon, F. (1945). Individual comparisons by ranking methods. *Biometrics Bulletin*, *1*(6), 80–83.
167. Wilcoxon, F. (1947). Probability tables for individual comparisons by ranking methods. *Biometrics*, *3*, 119–122.
168. Williams, J. S., & Lagakos, S. W. (1977). Models for censored survival analysis: Constant-sum and variable-sum models. *Biometrika*, *64*, 215–224.
169. Witkovsky, V. (2001). Computing the distribution of a linear combination of inverted gamma variables. *Kybernetika*, *37*(1), 79–90.
170. Witkovsky, V. (2002). Exact distribution of positive linear combinations of inverted Chi-square random variables with odd degrees of freedom. *Statistics & Probability Letters*, *56*, 45–50.
171. Wolfram, S. (2008). Approximate functions and interpolation. *Wolfram Mathematica Documentation Center*, Wolfram Research. http://reference.wolfram.com/mathematica/tutorial/.
172. Wolfram Software (2009). *Mathematica Version 7*.
173. Woolson, R. F. (2007). Wilcoxon signed-rank test. *Wiley encyclopedia of clinical trials* (Vol. 4, pp. 528–530). New York: Wiley.
174. Wu, J. W., Tsai, T., & Ouyang, L. (2001). Limited failure-censored life test for the Weibull distribution. *IEEE Transactions on Reliability*, *50*(1), 107–111.
175. Zellner, A. (1971). *An introduction to bayesian inference in econometrics*. New York: Wiley.
176. Zhang, L. F., Xie, M., & Tang, L. C. (2006). Robust regression using probability plots for estimating the weibull shape parameter. *Quality and Reliability Engineering International*, *22*(8), 905–917.
177. Zhang, L. F., Xie, M., & Tang, L. C. (2006). Bias correction for the least squares estimator of Weibull shape parameter with complete and censored data. *Reliability Engineering & System Safety*, *91*(8), 930–939.
178. Zheng, G., & Park, S. (2005). Another look at life testing. *Journal of Statistical Planning and Inference*, *127*, 103–117.

Index

A.G. Glen, L.M. Leemis (eds.), *Computational Probability Applications*, International Series in Operations Research & Management Science 247, DOI 10.1007/978-3-319-43317-2

B

C

MIX
Papier aus verantwortungsvollen Quellen
Paper from responsible sources
FSC® C105338
FSC
www.fsc.org

If you have any concerns about our products,
you can contact us on
ProductSafety@springernature.com

In case Publisher is established outside the EU,
the EU authorized representative is:
Springer Nature Customer Service Center GmbH
Europaplatz 3, 69115 Heidelberg, Germany

Printed by Libri Plureos GmbH
in Hamburg, Germany